Westermeier

Electrophoresis in Practice

Second Edition

VCH

A Wiley company

Reiner Westermeier

Electrophoresis in Practice

A Guide to Methods and Applications
of DNA and Protein Separations

Second Edition

in collaboration with

Jenny Fichmann
Sonja Gronau
Hanspeter Schickle
Günter Theßeling
Peter Wiesner

VCH
A Wiley company

Dr. Reiner Westermeier in collaboration with Jenny Fichmann
Pharmacia Biotech Europe GmbH   Sonja Gronau
Munzinger Straße 9   Hanspeter Schickle
D-79111 Freiburg   Günter Theßeling
Federal Republic of Germany   Peter Wiesner

Editorial Director: Dr. Hans-Joachim Kraus
Production Manager: Dipl.-Wirt.-Ing. (FH) H.-J. Schmitt

Cover illustration: Schneider, Raabe und Partner, Freiburg

Library of Congress Card No.: applied for

British Library Cataloguing-in-Publication Data:
A catalogue record for this book
is available from the British Library

Die Deutsche Bibliothek – CIP-Einheitsaufnahme:
Westermeier, Reiner:
Electrophoresis in practice : a guide to methods und
applications of DNA and protein separations / Reiner
Westermeier. In collab. with Neil Barnes ... - 2. ed. -
Weinheim : VCH, 1997
    Dt. Ausg. u.d.T.: Westermeier, Reiner: Elektrophorese-Praktikum
    ISBN 3-527-30070-8

© VCH Verlagsgesellschaft mbH, D-69451 Weinheim (Federal Republic of Germany), 1997

Printed on acid-free and chlorine-free paper

Composition: Kühn & Weyh, D-79015 Freiburg
Printing: betz-druck, D-64291 Darmstadt
Bookbinding: Großbuchbinderei J. Schäffer, D-67269 Grünstadt
Printed in the Federal Republic of Germany

Preface

The number of electrophoretic separation methods has increased dramatically since Tiselius' pioneer work for which he received the Nobel Prize. Development of these methods has progressed from paper, cellulose acetate membranes and starch gel electrophoresis to molecular sieve, disc, SDS, and immuno-electrophoresis and finally to isoelectric focusing but also to high resolution two-dimensional electrophoresis. Together with silver and gold staining, autoradiography, fluorography and blotting, these techniques afford better resolution, sensitivity and specificity for the analysis of proteins. In addition, gel electrophoresis has proved to be a unique method for DNA sequencing while high resolution two-dimensional electrophoresis has smoothed the fascinating path from isolation of the protein to the gene through amino acid sequencing and after gene cloning, to protein synthesis.

The spectrum of analytical possibilities has become so varied that an overview of electrophoretic separation methods seems desirable not only for beginners but also for experienced users. This book has been written for this purpose.

The author belongs to the circle of the *bluefingers* and experienced this in Milan in 1979 when he was accused of being a money forger when buying cigarettes in a kiosk after work because his hands were stained by Coomassie. Prof. Righetti and I had to extricate him from this tricky situation. According to Maurer's definition (Proceedings of the first small conference of the bluefingers, Tübingen 1972) an expert was at work on this book and he can teach the whitefingers, who only know of the methods by hearsay, for example, how not to get blue fingers.

As it is, I am sure that this complete survey of the methods will not only help the whitefingers but also the community of the bluefingers, silverfingers, goldfingers etc. and will teach them many technical details.

Weihenstephan, February 1990
Prof. Dr. Angelika Görg

Foreword

German version

This book was written for the practician in the electrophoresis laboratory. For this reason we have avoided physico-chemical derivations and formulas concerning electrophoretic phenomena.

The type of explanation and presentation stems from several years of experience in giving user seminars and courses, writing handbooks and solving user problems. They should be clear for technical assistants as well as for researchers in the laboratory. The commentary column offers room for personal notes.

In part I, an introduction - as short as possible - to the actual state of the art will be given. The references are not meant to be exhaustive.

Part II contains exact instructions for 11 chosen electrophoretic methods, which can be carried out with *one* single piece of equipment. The sequence of the methods was planned so that an electrophoresis course for beginners and advanced users can be established afterwards. The major methods used in biology, biochemistry, medicine and food science methods have been covered.

If - despite following the method precisely - unexplained effects should arise, their cause and the remedies can be found in the trouble-shooting guide in the appendix.

The author would be thankful for any additional comments and solutions for the trouble-shooting guide which the reader could supply.

Freiburg, March 1990   R. Westermeier

English version, First Edition

The author is grateful to Dr. Michael J. Dunn, Senior Lecturer at the National Heart and Lung Institute, Harefield, Middlesex, UK, for his kind engagement of reading the manuscript, correcting the english and for his excellent and informed advices.

In this version, some updates have been made to methodological aspects, new experiences, applications, and the references. A new drawing program is used, which allows higher resolution in the explanatory figures.

Leonberg, February 1993   R. Westermeier

English version, Second Edition

The author thanks Professor Görg for her tips for the state of the art of high resolution two-dimensional electrophoresis, and Dr. Gabriel Peltre, Institute Pasteur, Paris, for valuable hints on the practice of immunoelectrophoresis, agarose isoelectric focusing, blotting, and titration curves.

This version has been updated in the wording, the way of quoting the references, and in the methodology. A few figures, hints for problem solving, and a few very important references have been added. The main differences to the previous issue, however, are constituted by the addition of the lately developed methods for DNA typing and the methodology for vertical gels. Thus, section II contains now 15 chosen electrophoretic methods.

Freiburg, November 1996   R. Westermeier

Biography

Reiner Westermeier (born in 1951)

1976	Diplom Engineer
1981	Dr.-Ing., Technical University Munich, Weihenstephan, Germany
1976–1983	enganged in the development of new electro-phoresis systems and applications at the Technical University Munich
1984–1987	employed as an electrophoresis products and applications specialist at LKB Instrument GmbH and
1987–1990	at Pharmacia Biotech
1991	founded a new company called ETC Elektro-phorese-Technik for the development of new electrophoresis methods, media, and equipment
since January 1996	manager scientific support at Pharmacia Biotech in Freiburg, Germany
	several international cooperations and lectur-ing tours author of several publications and a book *(Electrophoresis in Practice*, First Edition, VCH)

Contents

Section I

Fundamentals

Introduction 1

1 Electrophoresis 5

1.0 General 5
1.1 Electrophoresis in non-restrictive gels 12
1.1.1 Agarose gel electrophoresis 12
1.1.2 Polyacrylamide gel electrophoresis of low-molecular weight substances 15
1.2 Electrophoresis in restrictive gels 16
1.2.1 The Ferguson plot 16
1.2.2 Agarose gel electrophoresis 17
1.2.3 Polyacrylamide gel electrophoresis of nucleic acids 19
       DNA sequencing 19
       DNA typing 22
       RNA and viroids 27
1.2.4 Polyacrylamide gel electrophoresis of proteins 28
       Disc electrophoreis 28
       Gradient gel electrophoresis 30
       SDS electrophoresis 31
       Two-dimensional electrophoresis 36

2 Isotachophoresis 41

3 Isoelectric focusing 45
3.1 Principles 45
3.2 Gels for IEF 47
3.3 Temperature 48
3.4 Controlling the pH gradient 48
3.5 The kinds of pH gradients 48
3.5.1 Free carrier ampholytes 48
3.5.2 Immobilized pH gradients 52
3.6 Preparative isoelectric focusing 55
3.7 Titration curve analysis 56

4 Blotting 59
4.1 Principle 59

4.2   Transfer methods   59
4.3   Blotting membranes   63
4.4   Buffers for electrophoretic transfers   64
4.5   General staining   66
4.6   Blocking   66
4.7   Specific detection   67
4.8   Protein sequencing   68
4.9   Transfer problems   69

5   **Instrumentation**   71
5.1   Current and voltage conditions   71
5.2   Power supply   73
5.3   Separation chambers   73
5.3.1   Vertical apparatus   67
5.3.2   Horizontal apparatus   75
5.4   Staining apparatus   77
5.5   Automated electrophoresis   77
5.6   Safety measures   79
5.7   Environmental aspects   79

6   **Interpretation of electropherograms**   81
6.1   Introduction   81
6.1.1   Purity control   81
6.1.2   Quantitative measurements   81
6.2   Computer aided analysis   83
6.2.1   Instrumentation for image aquisition   84
6.2.2   The optics of a densitometer   85
6.2.3   Integration and baseline   88

         Equipment for Section II   91
         Instrumentation   91
         Special laboratory equipment   93
         Consumables   94
         Chemicals   95

Section II

Methods

Method 1. PAGE of dyes 101
1   Sample preparation 101
2   Stock solutions 101
3   Preparing the casting cassette 101
4   Casting the ultrathin-layer gels 104
5   Electrophoretic separation 104

Method 2: Agarose and immunoelectrophoresis 107
1   Sample preparation 108
2   Stock solutions 108
3   Preparing the gels 108
4   Electrophoresis 112
5   Protein detection

Method 3: Titration curve analysis 119
1   Sample preparation 119
2   Stock solutions 119
3   Preparing the blank gels 120
4   Titration curve analysis 123
5   Coomassie and silver staining 126
6   Interpreting the curves 128

Method 4: Native PAGE in amphoteric buffers 131
1   Sample preparation 132
2   Stock solutions 132
3   Preparing the empty gels 133
4   Electrophoresis 137
5   Coomassie and silver staining 140

Method 5: Agarose IEF 143
1   Sample preparation 143
2   Preparing the agarose gel 144
3   Isoelectric focusing 146
5   Protein detection 148

Method 6: PAGIEF in rehydrated gels 151
1   Sample preparation 151
2   Stock solutions 152
3   Preparing the blank gels 152
4   Isoelectric focusing 155
5   Coomassie and silver staining 157
6   Densitometric evaluation 159
7   Perspectives 162

Method 7: SDS-polyacrylamide electrophoresis 165
1   Sample preparation 165
2   Stock solutions for the preparation of gels 169
3   Preparing the casting cassette 170
4   Gradient gel 172
5   Electrophoresis 176
6   Coomassie and silver staining 178
7   Blotting 181
8   Densitometry 181
9   Perspectives 184

Method 8: Semi-dry blotting of proteins 187
1   Transfer buffers 188
2   Technical execution 189
3   Staining of blotting membranes 193

Method 9: IEF in immobilized pH gradients 195
1   Sample preparation 196
2   Stock solutions 196
3   Immobiline recipes 197
4   Preparing the casting cassette 200
5   Preparing the pH gradient gels 201
6   Isoelectric focusing 208
7   Coomassie and silver staining 209
8   Strategies for IPG focusing 212

Method 10: High resolution 2D electrophoresis 213
1   Sample preparation 214
2   Stock solutions 215
3   Preparing the gel 216
4   Separation conditions 220
5   Coomassie and silver staining 223
6   Perspectives 228

Method 11: PAGE of double stranded DNA 229
1   Stock solutions 230
2   Preparing the gels 231
3   Sample preparation 234
4   Electrophoresis 235
5   Silver staining 240

Method 12: Native PAGE of single stranded DNA 243
1   Sample treatment 245
2   Gel properties 246
3   Buffers and additives 246
4   Conditions for electrophoresis 247
5   Strategies for SSCP analysis 248

Method 13: Denaturing gradient gel electrophoresis 249
1   Sample preparation 250
2   Rehydration solutions 250
3   Preparing the rehydration cassette 250
4   Rehydration 252
5   Electrophoresis 254
6   Silver staining 256

Method 14: Denaturing PAGE of DNA 257
1   Sample preparation 258
2   Solutions 258
3   Rehydration 259
4   Electrophoresis 259
5   Silver staining 262

Method 15: Vertical PAGE 263
1   Sample preparation 264
2   Stock solutions 264
3   Single gel casting 265
4   Multiple gel casting 268
5   Electrophoresis 272
6   SDS electrophoresis of small peptides 273
7   Two-dimensional electrophoresis 275
8   DNA electrophoresis 276
9   Long shelflife gels 276
10 Coomassie and silver staining 276

Appendix

A Trouble-shooting guide 277

A1   Isoelectric focusing 277
A1.1 PAGIEF with carrier ampholytes 277
A1.2 Agarose IEF with carrier ampholytes 285
A1.3 Immobilized pH gradients 288
A2   SDS-electrophoresis 294
A3   Semi-dry blotting 302
A4   Two-dimensional electrophoresis (IPG-DALT) 308
A5   DNA electrophoresis 312
A6   Vertical PAGE 315

References 317

Index 323

Abbreviations, symbols, units

A	Ampere
acc.	according
A,C,G,T	Adenine, cytosine, guanine, thymine
ACES	N-2-acetamido-2-aminoethanesulfonic acid
A/D-transformer	Analog-digital transformer
AFLP	Amplified restriction fragment length polymorphism
APS	Ammonium persulphate
ARDRA	Amplified ribosomal DNA restriction analysis
AU	Absorbance units
BAC	Bisacryloylcystamine
Bis	N, N'-methylenebisacrylamide
bp	Base pair
BSA	Bovine serum albumin
C	Crosslinking factor [%]
CAPS	3-(cyclohexylamino)-propanesulfonic acid
CDGE	Constant denaturing gel electrophoresis
CE	Capillary electrophoresis
CHAPS	3-(3-cholamidopropyl)dimethylammonio-1-propane sulfonate
CM	Carboxylmethyl
CMW	Collagen molecular weight
concd	Concentrated
const.	Constant
CTAB	Cetyltrimethylammonium bromide
Da	Dalton
DAF	DNA amplification fingerprinting
DBM	Diazobenzyloxymethyl
DDRT	Differential display reverse transcription
DEAE	Diethylaminoethyl
DGGE	Denaturing gradient gel electrophoresis
Disc	Discontinuous
DMSO	Dimethylsulfoxide
DNA	Desoxyribonucleic acid
DPT	Diazophenylthioether
DSCP	Double strand conformation polymorphism
dsDNA	Double stranded DNA
DTE	Dithioerythreitol
DTT	Dithiothreitol
E	Field strength in V/cm
EDTA	Ethylenediaminetetraacetic acid
ESI	Electro spray ionization

GC	Group specific component
h	Hour
HEPES	N-2-hydroxyethylpiperazine-N'-2-ethananesulfonic acid
HMW	High Molecular Weight
HPCE	High Performance Capillary Electrophoresis
HPLC	High Performance Liquid Chromatography
I	Current in A, mA
IEF	Isoelectric focusing
IgG	Immunoglobulin G
IPG	Immobilized pH gradients
IPG-Dalt	2D electrophoresis: IPG/SDS electrophoresis
Iso-Dalt	2D electrophoresis: IEF/SDS electrophoresis
ITP	Isotachophoresis
kB	Kilobases
kDa	Kilodaltons
K_R	Retardation coefficient
LDAO	Lauryldimethylamine-N-oxide
LMW	Low Molecular Weight
mA	Milliampere
MALDI	Matrix assisted laser desorption ionization
MEKC	Micellar electrokinetic chromatography
MES	2-(N-morpholino)ethanesulfonic acid
min	Minute
mol/L	Molecular mass
MOPS	3-(N-morpholino)propanesulfonic acid
m_r	Relative electrophoretic mobility
mRNA	messenger RNA
MW	Molecular weight
NAP	Nucleic Acid Purifier
Nonidet	Non-ionic detergent
O.D.	Optical density
P	Power in W
PAG	Polyacrylamide gel
PAGE	Polyacrylamide gel electrophoresis
PAGIEF	Polyacrylamide gel isoelectric focusing
PBS	Phosphate buffered saline
PCR®	Polymerase Chain Reaction
PEG	Polyethylene glycol

PFG	Pulsed Field Gel (electrophoresis)
PGM	Phosphoglucose mutase
pI	Isoelectric point
PI	Protease inhibitor
pK value	Dissociation constant
PMSF	Phenylmethyl-sulfonyl fluoride
PVC	Polyvinylchloride
PVDF	Polyvinylidene difluoride
r	Molecular radius
RAPD	Random amplified polymorphic DNA
REN	Rapid efficient nonradioactive
RFLP	Restriction fragment length polymorphism
R_f value	Relative distance of migration
R_m	Relative electrophoretic mobility
RNA	Ribonucleic acid
RPA	Ribonuclease protection assay
s	Second
SDS	Sodium dodecyl sulfate
ssDNA	single stranded DNA
T	Total acrylamide concentration [%]
t	Time, in h, min ,s
TBE	Tris borate EDTA
TCA	Trichloro acetic acid
TEMED	N,N,N',N'-tetramethylethylenediamine
TF	Transferrin
TGGE	Temperature gradient gel electrophoresis
TMPTMA	Trimethylolpropane-trimethacrylate[2-ethyl-2(hydrohymethyl)R 1,3-propandiol-trimethacrylate]
Tricine	N,tris(hydroxymethyl)-methyl glycine
Tris	Tris(hydroxymethyl)-aminoethane
U	Volt
V	Volume in L
v	Speed of migation in m/s
VLDL	Very low density lipoproteins
v/v	Volume per volume
W	Watt
w/v	Weight per volume (mass concentration)
ZE	Zone electrophoresis
2D electrophoresis	Two-dimensional electrophoresis

Introduction

For no other biochemical separation method nowadays does one find so many new developments and methods as for electrophoretic separation techniques. With this method a high separation efficiency can be achieved using a relatively limited amount of equipment. The main fields of application are biological and biochemical research, protein chemistry, pharmacology, forensic medicine, clinical investigations, veterinary science, food control as well as molecular biology. It will become increasingly important to be able to choose and carry out the appropriate electrophoresis technique for specific separation problems.

The monograph by Andrews (Andrews 1986) is one of the most complete and practice-oriented books about electrophoretic methods. In the present book, electrophoretic methods and their applications will be presented in a much more condensed form.

Andrews AT. Electrophoresis, theory techniques and biochemical and clinical applications. Clarendon Press, Oxford (1986).

Principle: Under the influence of an electrical field, charged molecules and particles migrate in the direction of the electrode bearing the opposite charge. During this process, the substances are in aqueous solution. Because of their varying charges and masses, different molecules and particles of a mixture will migrate at different speeds and will thus be separated into single fractions.

The electrophoretic mobility which influences the speed of migration, is a significant and characteristic parameter of a charged molecule or particle and is dependent on the pK value of the charged group and the size of the molecule or particle. It is influenced by the type, concentration and pH of the buffer, by the temperature and the field strength as well as by the nature of the support material. Electrophoretic separations can be carried out in free solutions as in capillary electrophoresis or systems without support phases but also in stabilizing media such as thin-layer plates, films or gels.

Chrambach A. The practice of quantitative gel electrophoresis. VCH Weinheim (1985).

Mosher RA, Saville DA, Thormann W. The Dynamics of Electrophoresis. VCH Weinheim (1992).

Detailed theoretical explanations can be found in the books by Chrambach (1985) and Mosher *et al.* (1992).

The *relative* electrophoretic mobility of substances is usually specified. It is calculated relative to the migration distance of a standard substance applied in the same run so as to compensate for different field strengths and separation time.

The relative mobility is abbreviated as mr or Rm.

Three basically different electrophoretic separation methods exist:

a) Electrophoresis, sometimes called zone electrophoresis (ZE).
b) Isotachophoresis or ITP
c) Isoelectric focusing or IEF

The three separation principles are illustrated in Fig. 1.

Electrophoresis is a general term for the three methods. Blotting is not a separation but a detection method.

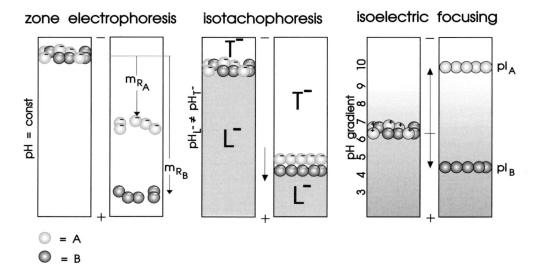

Fig. 1: The three electrophoretic separation principles. Explanations in the text. A and B are the components of the sample.

a) In *zone electrophoresis* a homogeneous buffer system is used over the whole separation time and range so as to ensure a constant pH value. The migration distances during a defined time limit are a measure of the electrophoretic mobility of the various substances.

This is also valid for disc electrophoresis, a discontinuous system exists only at the beginning of the separation and changes into a homogeneous one.

b) In *isotachophoresis* (ITP), the separation is carried out in a discontinuous buffer system. The ionized sample migrates between a leading electrolyte with a high mobility and a terminating ion with a low mobility, all of them migrating with the same speed. The different components are separated according to their electrophoretic mobilities and form stacks: the substance with the highest mobility directly follows the leading ion, the one with the lowest mobility migrates directly in front of the terminating electrolyte. This method can be used for quantitative analysis.

In comparison to other electrophoretic and chromatographic separation methods, ITP is considered exotic because there are no spaces between the zones: the bands are not "peaks" (Gaussian curves) but "spikes" (concentration dependent bands).
ITP is mostly applied for stacking of the samples during the first phase of disc electrophoresis.

c) *Isoelectric focusing* (IEF) takes place in a pH gradient and can only be used for amphoteric substances such as peptides and proteins. The molecules move towards the anode or the cathode until they reach a position in the pH gradient where their net charges are zero. This pH value is the *"isoelectric point"* (pI) of the substance. Since it is no longer charged, the electric field does not have any influence on it. Should the substance diffuse away, it will gain a net charge again, and the applied electric field will cause it to migrate back to its pI. This concentrating effect leads to the name *focusing*.

In IEF it is important to find the correct place in the pH gradient to apply the sample, since some substances are unstable at certain pH values (see below).

Area of application: Electrophoretic methods are used for the qualitative characterization of a substance or mixture of substances, for control of purity, quantitative determinations, and preparative purposes.

The scope of the applications ranges from whole cells and particles to nucleic acids, proteins, peptides, amino acids, organic acids and bases, drugs, pesticides and inorganic anions and cations – in short everything that can carry a charge.

The sample: An important criterion for the choice of the appropriate electrophoretic method is the nature of the sample to be analyzed. There must be no solid particles or fatty components suspended in the solution. Those interfere with the separation by blocking the pores of the matrix. Sample solutions are mostly centrifuged before electrophoresis.

Sample application on gels which are immersed in buffer (e.g. vertical and submarine gels) is done with syringes into sample wells polymerized into the gel or into glass tubes, with the sample made denser than the buffer with glycerol or sucrose.

The easiest is the separation of substances which are exclusively negatively or positively charged:
Examples of such anions or cations are: nucleic acids, dyes, phenols and organic acids or bases. Amphoteric molecules such as amino acids, peptides, proteins and enzymes have a positive or negative net charge depending on the pH of the buffer, because they possess acidic as well as basic groups.

Macromolecules such as proteins and enzymes are sometimes sensitive to certain pH values or buffer substances; conformational changes, denaturation, complex formation, and intermolecular interactions can also occur. The concentration of the substances in the solution also plays a role. In particular, when the sample enters the gel, overloading effects can occur when the protein concentration reaches a critical value during the transition from the solution into the more restrictive gel matrix.

For sodium dodecyl sulphate electrophoresis, the sample must first be denatured; which means it must be converted into molecule-detergent micelles. The method of selective sample extraction, particularly the extraction of not easily soluble substances often determines the nature of the buffer to be used.

The nature of the stabilizing medium, e.g. a gel, is dependent on the size of the molecule to be analyzed.

For open surfaces as in horizontal systems (e.g. cellulose acetate, agarose gels and automated electrophoresis) sample applicators are used or else the sample is pipetted into sample well with a micropipette. For capillary systems syringes are also used, though most instruments have an automatic sample applicator.

The buffer: The electrophoretic separation of samples is done in a buffer with a precise pH value and a constant ionic strength. The ionic strength should be as low as possible so that both the contribution of the sample ions to the total current and their speed will be high enough.

During electrophoresis, the buffer ions are carried through the gel just like the sample ion: negatively charged ions towards the anode, positively charged ones towards the cathode. This should be achieved with as little energy as possible so that not much Joule heat is developed. Yet a minimum buffering capacity is required so that the pH value of the samples analyzed does not have any influence on the system.

To guarantee constant pH and buffer conditions the supplies of electrode buffers must be large enough. The use of buffer gel strips or wicks instead of tanks is very practical, though only feasible in horizontal systems.

For anionic electrophoresis very basic, and for cationic electrophoresis very acidic buffers are used.

In vertical or capillary systems, the pH is very often set to a very high (or low) value, so that as many as possible sample molecules are negatively (or positively) charged, and thus migrate in the same direction.

In these systems the sample is loaded at one end of the separation medium.

When a very clean gel is used, also amphoteric buffers can be applied, which do not migrate themselves during electrophoresis. Such a buffer substance must, however, possess a high buffering capacity at its isoelectric point. For some applications, no buffer reservoirs are necessesary with this method.

See method 5 in this book.

Electroendosmosis: The static support, the stabilizing medium (e.g. the gel) and/or the surface of the separation equipment such as glass plates, tubes or capillaries can carry charged groups: e.g. carboxylic groups in starch, sulfonic groups in agarose. These groups become ionized in basic and neutral buffers: in the electric field they will be attracted by the anode. As they are fixed in the matrix, they can not migrate. This results in a compensation by the counterflow of H_3O^+ ions towards the cathode: electroendosmosis. This effect is observed as a water flow towards the cathode, which carries the solubilized substances along. The electrophoretic and electroosmotic migrations are then additive.

Electroendosmosis generally interferes with electrophoretic separations, yet a few methods take advantage of this effect to achieve separation or detection results (see page 7: MEKC and page 14: counter immunoelectrophoresis).

In capillary electrophoresis mostly the term "electroosmotic flow" is applied, the term "electroendosmosis" is only used in gel electrophoresis.

1 Electrophoresis

1.0 General

Electrophoresis in free solution

Moving boundary electrophoresis: Arne Tiselius (1937) developed the moving boundary technique for the electrophoretic separation of substances, for which, besides his work on adsorption analysis, he received the Nobel prize in 1948.

The sample, a mixture of proteins for example, is applied in a U-shaped cell filled with a buffer solution and at the end of which electrodes are immersed. Under the influence of the applied voltage, the compounds will migrate at different velocities towards the anode or the cathode depending on their charges. The changes in the refractive index at the boundary during migration can be detected at both ends on the solution using Schlieren optics.

Tiselius A. Trans Faraday Soc. 33 (1937) 524-531.

Nowadays moving boundary electrophoresis in free solution is mainly used in fundamental research to determine exact electrophoretic mobilities.

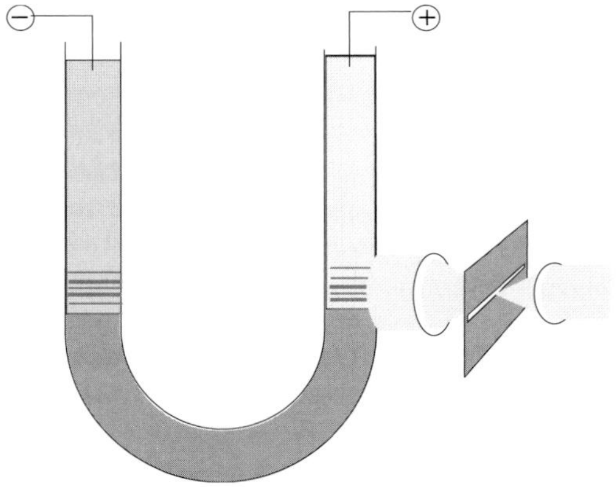

Fig. 2: Moving boundary electrophoresis in a U-shaped cell according to Tiselius. Measurement of the electrophoretic mobility with Schlieren optics.

Free flow electrophoresis: in this technique developed by Hannig (1982) a continuous stream of buffer flows perpendicular to the electrical field through a buffer film between two cooled glass plates which is 0.5 to 1.0 mm wide. At one end the sample is injected at a defined spot and at the other end, the fractions are collected in an array of tubes.

The varying electrophoretic mobilities perpendicular to the flow lead to differently heavy but constant deviations of the components so that they reach the end of the separation chamber at different though stable positions (see Fig. 3).

Hannig K. Electrophoresis. 3 (1982) 235-243.

This is the only continuous electrophoretic separation method.

Wagner H, Kuhn R, Hofstetter S. In: Wagner H, Blasius E. Ed. Praxis der elektrophoretischen Trennmethoden. Springer-Verlag, Heidelberg (1989) 223-261.

Besides the separation of soluble substances, this technique is also used for the identification, purification and isolation of cell organelles and membranes or whole cells such as erythrocytes, leukocytes, tissue cells, the causal agent of malaria and other parasites (Hannig 1982, Wagner et al. 1989).

This method is very effective since even minimal differences in the surface charge of particles and cells can be used for separation.

Unfortunately electrophoresis in free solution cannot yet be applied on an industrial scale. The upscaling of the instrumentation is limited by the thermal convection which results from the insufficient dissipation of Joule heat from the flowing electrolyte. Loading cannot be freely increased because highly concentrated samples begin to sediment. Both these limiting factors occur only under gravity. Since 1971, ever since Apollo 14, experiments have been conducted in space to try and develop production in an orbital station.

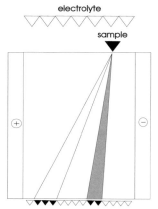

Fig. 3: Schematic drawing of a continuous free flow electrophoresis system. According to Wagner et al. (1989)

Capillary electrophoresis(CE): this technique is being used increasingly for analytical and micropreparative electrophoresis (Jorgenson and Lukacs, 1981; Hjertén, 1983): as for HPLC, the abbreviation HPCE for *High Performance Capillary Electrophoresis* is often used. Separation is carried out in a fused silica capillary 20 to 30 cm long and with an internal diameter of 50 to 100 μm. Both ends are immersed in a buffer container into which the electrodes are built.

Jorgenson JW, Lukacs KD. AnalChem 53 (1981) 1298-1302 Hjertén S. J Chromatogr. 270 (1983) 1-6.

Fused silica capillaries are otherwise used in gas chromatography.

The amount of chemicals and sample needed is very low. The volume of injected material is usually not more than 2 − 4 nL, nanograms of sample material is required.

Field strengths of up to 1 kV/cm and currents of 10 to 20 mA are used; for this reason a power supply which can yield voltages up to 30 kV is needed. Joule heat can be dissipated very effectively from these thin capillaries with a fan.

CE separations typically take 10 to 20 min. There are many detection methods possible: UV/VIS, fluorescence, conductivity, electrochemistry etc. In most applications the fractions are detected by UV measurement at 280, 260 or in some cases even 185 nm directly in the capillary.

For some substances and applications the limit of detection can go as low as to the attomole level.

In general the results are then further processed by HPLC interpretation software on personal computers.

To prevent adsorption of components on the surface of the capillary and electro-osmotic effects, the inside of the capillary can be coated with linear polyacrylamide or methyl cellulose. Capillary electrophoresis instruments can be used for all three of the separation methods: electrophoresis, isotachophoresis and isoelectric focusing. Even an additional new method, a hybrid of electrophoresis and chromatography, has been developed:

Micellar electrokinetic chromatography (MEKC) introduced by Terabe et al. (1984). It is the only electrophoretic method, which can separate neutral as well as charged compounds. Surfactants are used at concentrations over the the critical micelle concentration. The charged micelles migrate in the opposite direction to the electro-osmotic flow created by the capillary wall. The electro-osmotic counter flow is faster than the migration of the micelles. During migration, the micelles interact with the sample compounds in a chromatographic manner through both hydrophobic and electrostatic interactions.

It has become one of the most widely used CE methods. More details on this method are found in a review by Terabe et al. (1994).

The buffer used depends on the nature of the separation: e.g. 20 to 30 mmol/L sodium phosphate buffer pH 2.6 for electrophoresis of peptides.

*Terabe S, Otsuka K, Ichikawa K, Tsuchiya A, Ando T. Anal-Chem.64 (1984) 111-113
Terabe S, Chen N, Otsuka K. In Chrambach A, Dunn M, Radola BJ. Eds. Advances in Electrophoresis 7. VCH Weinheim (1994) 87-153.*

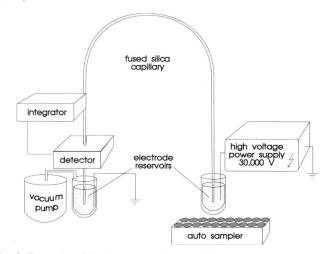

Fig. 4: Example of the instrumentation used for capillary electrophoresis.

One great advantage of capillary electrophoresis lies in its automation. Every step can be controlled by semiautomatic or full automatic instrumentation. An autosampler is a standard part of this equipment.

Another advantage is the possibility of linking with other analytical instruments either before electrophoresis: HPLC/HPCE or after: HPCE/MS.

However, the investment for such an instrument is by far higher compared to a gel electrophoresis equipment.

For preparative separations a fraction collector is attached to the UV detector. The identification of the individual substances is done by the relative mobility or the molecular weight, or else the collected fractions are analyzed.

In contrast to Reversed Phase Chromatography proteins are not damaged during HPCE and, in addition, the resolution is better.

For molecular weight separations of proteins, peptides,and nucleic acids capillaries filled with polyacrylamide gel are used (Cohen et al. 1987).

Cohen AS, Karger BL. J Chromatogr. 397 (1987) 409-417.

Electrophoresis in supporting media

Compact material such as paper, films or gels are used. So as to monitor the progress of the separation and to recognize the end of the run, dyes with a high electrophoretic mobility are applied together with the sample.

The instructions in the second part are limited to electrophoresis in supporting media since these techniques only require minimal equipment.

For separation of proteins in anodal direction Bromophenol Blue, Xylenecyanol or Orange G are used, in the cathodal direction Bromocresol Green, Pyronine or Methylene Blue.

Detection of the separated zones can either be done directly in the medium by staining with Coomassie blue or silver staining, by spraying with specific reagents, by enzyme substrate coupling reactions, immunoprecipitation, autoradiography, fluorography or indirectly by immunoprinting or blotting methods. Recently, a comprehensive survey on enzyme electrophoresis and specific staining methods has been published by Rothe (1994).

Blotting: transfer to immobilizing membranes followed by staining or specific ligand binding.

Rothe G. Electrophoresis of Enzymes. Springer Verlag, Berlin, (1994).

Paper and thin-layer electrophoresis: These methods have mostly been abandoned in profit of gel electrophoresis, because of improved separation and the higher loading capacity of agarose and polyacrylamide gels. Electrophoretic separations on thin-layer silica gel plates linked to buffer tanks are only carried out for the analysis of polysaccharides of high molecular weight and lipopolysaccharides, which can obstruct the pores of the gels (Scherz, 1990).

Scherz H. Electrophoresis. 11 (1990) 18-22.

Cellulose acetate membrane electrophoresis: Cellulose acetate membranes have large pores and therefore hardly exert any sieving effect on proteins (Kohn, 1957). This means that electrophoretic separations are entirely based on charge density. The matrix exerts little effect on diffusion so that the separated zones are relatively wide while the resolution and limit of detection are low. On the other hand they are easy to handle and separation and staining are rapid. The cellulose acetate strips are suspended in the tank of a horizontal apparatus, so that both ends dip in the buffer; no cooling is necessary during separation. This technique is widely used for routine clinical analysis and related applications for the analysis of serum or isoenzymes.

Kohn J. Nature 180 (1957) 986-988

Because the resolution and reproducibility of separations in agarose and polyacrylamide gels are better, cellulose acetate membranes are more and more often replaced by gel electrophoresis.

Gel electrophoresis

The gel: The gel matrix should have adjustable and regular pore sizes, be chemically inert and not exhibit electroosmosis. Vertical cylindrical gel rods or plates as well as horizontal gel slabs are available, the latter being usually cast on to stable support film to facilitate handling (Fig. 5).

The instructions in the second part are limited to horizontal gels on support films since these can be used for all methods and with universally applicable equipment.

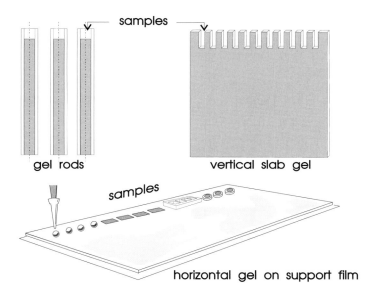

Fig. 5: Gel geometries for electrophoretic separations.

Starch gels were introduced by Smithies (1955) and are prepared from hydrolyzed potato starch which is dissolved by heating and poured to a thickness of 5 to 10 mm. The pore size can be adjusted by the starch concentration of the solution. Because of the low reproducibility and the impractical handling these gels have been largely replaced by polyacrylamide gels.

Smithies O. Biochem J. 61 (1955) 629-641.

Starch is a natural product whose properties can vary greatly.

Agarose gels are mostly used when large pores for the analysis of molecules over 10 nm in diameter are needed. Agarose is a polysaccharide obtained from red seaweed.

By removal of the agaropectin, gels of varying electroosmosis and degrees of purity can be obtained. They are characterized by their melting point (35 °C to 95 °C) and the degree of electro-endosmosis (m_r).

m_r is dependent on the number of polar groups left. The definition is the same like for relative electrophoretic mobility.

The pore size depends on the concentration of agarose: one usually refers to the weight of agarose and the volume of water. The unavoidable losses of water which occur during heating can vary from batch to batch, so in practice, this value cannot be absolutely exact. In general gels with a pore size from 150 nm at 1% (w/v) to 500 nm at 0.16% are used.

For pore diameters up to 800 nm (0.075% agarose): Serwer P. Biochemistry. 19 (1980) 3001-3005.

Agarose is dissolved in boiling water and then forms a gel upon cooling. During this process double helixes form which are joined laterally to form relatively thick filaments (Fig. 6).

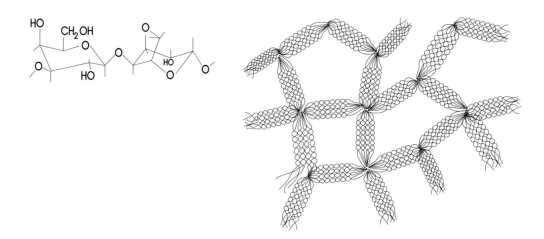

Fig. 6: Chemical structure of agarose and structure of the polymers during gel formation.

For DNA separations 1 to 10 mm thick gels are cast on UV-transparent trays, because the bands are usually stained with fluorescent dyes: Ethidium bromide or SYBR Green.

The gels are run under buffer in order to prevent drying out due to electroendosmosis.

For protein electrophoresis the gels are made by coating horizontal glass plates or support films with a solution of agarose. The thickness of the gel is determined by the volume of the solution and the surface it covers.

Very even gel thicknesses are obtained by pouring the solution in prewarmed molds.

Polyacrylamide gels, which were first used for electrophoresis by Raymond and Weintraub (1959), are chemically inert and particularly mechanically stable. By chemical co-polymerization of acrylamide monomers with a cross-linking reagent – usually N,N'-methylenebisacrylamide (Fig. 7) – a clear transparent gel which exhibits very little electroendosmosis is obtained.

Raymond S. Weintraub L. Science. 130 (1959) 711-711. The reaction is started with ammonium persulphate as catalyst, TEMED provides the tertiary amino groups to release the radicals.

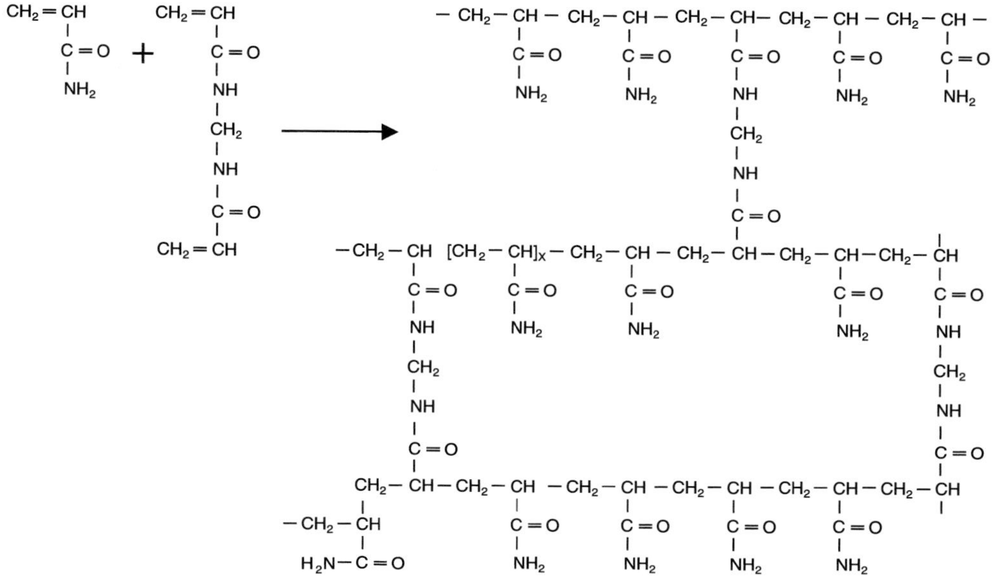

Fig. 7: The polymerization reaction of acrylamide and methylenebisacrylamide.

The pore size can be exactly and reproducibly controlled by the total acrylamide concentration T and the degree of cross-linking C (Hjertén, 1962):

Hjertén S.Arch Biochem Biophys Suppl 1 (1962) 147.

$$T = \frac{(a+b) \times 100}{V} \ [\%], \quad C = \frac{b \times 100}{a+b} \ [\%]$$

a is the mass of acrylamide in g,
b the mass of methylenebisacrylamide in g, and
V the volume in mL.

When C remains constant and T increases, the pore size decreases. When T remains constant and C increases, the pore size follows a parabolic function: at high and low values of C the pores are large, the minimum being at $C=4\%$.

Gels with C > 5 % are brittle and relatively hydrophobic. They are only used in special cases.

Besides methylenebisacrylamide a number of other cross-linking reagents exist, they have been listed and compared by Righetti (1983). N,N'-bisacryloylcystamine will be mentioned here, it possesses a disulfide bond which can be cleaved by thiol reagents. Because of this, it is possible to solubilize the gel matrix after electrophoresis.

Righetti PG.: Isoelectric focusing: theory, methodology and applications. Elsevier Biomedical Press, Amsterdam (1983).

Polymerization should take place under an inert atmosphere since oxygen can act as a free radical trap. The polymerization is temperature dependent: to prevent incomplete polymerization the temperature should be maintained above 20 °C.

The monomers are toxic and should be handled with precaution.

To minimize oxygen absorption gels are usually polymerized in vertical casting chambers: cylindrical gels in glass tubes and flat gels in moulds formed by two glass plates sealed together around the edges.

Besides the vertical casting techniques, horizontal ones also exist. The increased oxygen intake must be compensated by a higher amount of catalyst. This can lead to problems during separation.

For electrophoresis in vertical systems the gel in glass rods or cassettes are put in the buffer tanks, and are in direct contact with the electrode buffers. Gels for horizontal systems are polymerized on a support and removed from the mould before use.

For sample application wells are formed at the upper edge of the gel during polymerization (see Fig. 5). These are made by insertion of a sample comb between the glass plates. In horizontal gels, sample wells are not always necessary; the samples can be applied directly on the surface with strips of filter paper or silicone rubber.

In homogeneous buffer systems, narrow sample slots on the surface of horizontal gels are also important to obtain good results.

The various gel electrophoresis methods can be divided into those in restrictive and non-restrictive media. Restrictive gels work against diffusion so the zones are more distinctly separated and better resolved than in non-restrictive gels. The limit of detection is thus increased.

In restrictive gels, the size of the molecule has a major influence on the result of the separation.

1.1 Electrophoresis in non-restrictive gels

For these techniques the frictional resistance of the gel is kept negligibly low so that the electrophoretic mobility depends only on the net charge of the sample molecule. Horizontal agarose gels are used for high molecular weight samples such as proteins or enzymes and polyacrylamide gels for low molecular weight peptides or polyphenols.

1.1.1 Agarose gel electrophoresis

Zone electrophoresis

Agarose gels with concentrations of 0.7 to 1% are often used in clinical laboratories for the analysis of serum proteins. The separation times are exceedingly low: about 30 min. Agarose gels are also used for the analysis of isoenzymes of diagnostic importance such as lactate dehydrogenase (Fig. 8) and creatine kinase.

Because of their large pore size, agarose gels are especially suited to specific protein detection by *immunofixation:* after electrophoresis the specific antibody is allowed to diffuse through the gel. The insoluble immunocomplexes formed with the respective antigen result in insoluble precipitates and the non-precipitated proteins can be washed out. In this way only the desired fractions are detected during development.

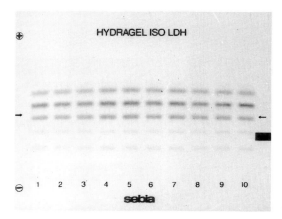

Fig. 8: Agarose electrophoresis of lactate dehydrogenase isoenzymes. Specific staining with the zymogram technique.

Immunoprinting functions in a similar way: after the electrophoretic separation, an agarose gel containing antibodies or a cellulose acetate membrane impregnated with antibodies is placed on the gel. The antigens then diffuse towards the antibodies and the identification of the zone is done in the antibody-containing medium. Immunoprinting is mainly used for gels with small pores.

Besides immunofixing and immunoprinting, immunoblotting also exists for protein identification : immobilizing membranes, for example nitrocellulose, are used on the surface of which the proteins are adsorbed, see "blotting" on page 59 ff.

Immunoelectrophoresis

The principle of immunoelectrophoresis is the formation of precipitate lines at the equivalence point of the antigen and its corresponding antibody. In this method it is important that the ratio between the quantities of antigen and antibody be correct (antibody titer). When the antibody is in excess, statistically at most one antigen binds to each antibody while when the antigen is in excess at most one antibody binds to each antigen. Yet at a specific antigen/antibody ratio (equivalence point) huge macromolecules are formed. They consist of an antigen-antibody-antigen-antibody-... sequence and are immobilized in the gel matrix as an immunoprecipitate. The white precipitate lines are visible in the gel and can be revealed with protein stains. The method is specific and the sensitivity very high because distinct zones are formed. Immunoelectrophoresis can be divided into three principles (Fig. 9):

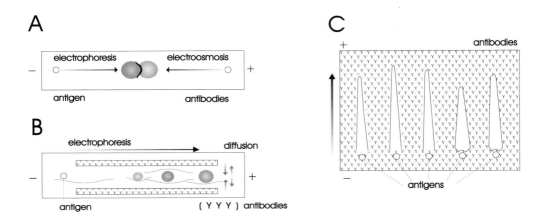

Fig. 9: The three principles of immunoelectrophoresis A, B and C, see text for details.

A. *Counter immunoelectrophoresis* according to Bussard (1959): in an agarose gel exhibiting high electro-osmosis, the buffer is set at a pH about 8.6 so that the antibody does not carry any net charge. The sample and the antibody are placed in their respective wells and move towards each other: the charged antigens migrate electrophoretically and the antibodies are carried by the electro-osmotic flow.

Bussard A. Biochim. Biophys Acta. 34 (1959) 258-260.

B. *Zone electrophoresis/immunodiffusion* according to Grabar and Williams (1953): first a zone electrophoresis is run in an agarose gel, followed by the diffusion of the antigen fraction towards the antibody which is pipetted into troughs cut in the side parallel to the electrophoretic run.

Grabar P, Williams CA. Biochim Biophys Acta. 10 (1953) 193.

C. The *"rocket"* technique according to Laurell (1966) and the related methods: antigens migrate in an agarose gel which contains a definite concentration of antibody. As in method A the antibodies are not charged because of the choice of the buffer. As the sample migrates one antibody will bind to one antigen until the ratio of concentrations corresponds to the equivalence point of the immunocomplex.
    The result is that rocket shaped precipitation lines are formed,

Laurell CB. Anal Biochem. 15 (1966) 45-52.

the enclosed areas are proportional to the concentration of antigen ion in the sample. A series of modifications to this technique exist, including two-dimensional ones.

Affinity electrophoresis

This is a method related to immunoelectrophoresis which is based on the interactions between various macromolecules for example lectin-glycoprotein, enzyme-substrate and enzyme-inhibitor complexes (Bøg-Hansen and Hau, 1981)

Bøg-Hansen TC, Hau J. J Chrom Library. 18 B (1981) 219-252.

All the known immunoelectrophoretic techniques are used here. For example, specific binding lectin collected worldwide from plant seeds are examined with line affinity electrophoresis. In this way carbohydrate changes in glycoproteins during different biological processes can be identified. In Fig. 10 an application of affinity electrophoresis to differentiate between alkaline phosphatase of liver and bone is shown.

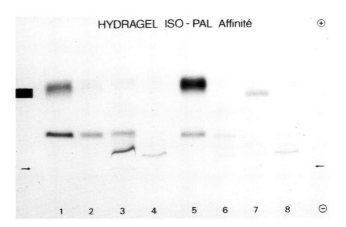

HYDRAGEL ISO - PAL Affinité   ⊕

Fig. 10: Affinity electrophoresis of isoenzymes of alkaline phosphatase from the liver and the bones. The wheat germ agglutinin specifically binds the bone fraction which is recognizable as a characteristic band close to the application point.

1.1.2 Polyacrylamide gel electrophoresis of low molecular weight substances

Since low molecular weight fractions cannot be chemically fixed in a matrix with large pores, horizontal polyacrylamide gels polymerized on ultra-thin films are used. They are dried at 100 °C immediately after electrophoresis and then sprayed with specific reagents. With this method for example, dyes with molecular weights of approximately 500 Da* can be separated.

*See method 1
According to the guide-lines of the SI, the use of the term Dalton for 1.6601 x 10⁻²⁷ kg is no longer recommended. However it is still a current unit in biochemistry.

1.2 Electrophoresis in restrictive gels

1.2.1 The Ferguson plot

Although during electrophoresis in restrictive gels, electrophoretic mobility depends both on net charge and on molecular radius this method can also be used for the physico-chemical analysis of proteins. The principle was formulated by Ferguson (1964): the samples are separated under identical buffer, time and temperature conditions but with different gel concentrations (g/100 mL for agarose, $\% T$ for polyacrylamide). The distances traveled will vary: m_r is the relative mobility. A plot of log 10 m_r versus the gel concentration yields a straight line.

Ferguson KA. Metabolism. 13 (1964) 985-995.

The slope (see Fig. 11) is a measure of the molecular size and is called the retardation coefficient K_R.

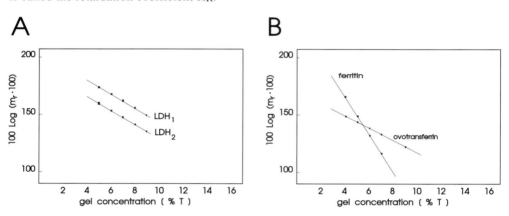

Fig. 11: Ferguson plots: plots of the electrophoretic migrations of proteins versus gel concentrations. (A) Lactate dehydrogenase isoenzymes; (B) differing proteins. See text for further details.

For globular proteins there is a linear relationship between K_R and the molecular radius r (Stokes radius), so the molecular size can be calculated from the slope of the plot. Once the free mobility and the molecular radius are known the net charge can also be calculated (Hedrick and Smith, 1968). For protein mixtures the following deductions can be made according to the appearance of the plots:

Hedrick JL, Smith AJ. Arch Biochem Biophys. 126 (1968) 155-163.

● The lines are parallel: The proteins have the same size but different mobilities e.g. iosenzymes.

Fig. 11A

● The slopes are different but the lines do not cross: the protein corresponding to the upper curve is smaller and has a higher net charge.

● The lines cross beyond *T*=2%: the larger protein has the higher charge density and intercepts the y-axis at a higher value (Fig. 11B).

Fig. 11B

● Several lines cross at a point where *T*< 2%: these are obviously the various polymers of one protein.

Same net charge, different molecular sizes.

1.2.2 Agarose gel electrophoresis

Proteins

Since highly concentrated agarose gels above 1% (1g/100 mL agarose in water) are cloudy and the electro-osmotic flow is high, agarose gels are only used for the separation of very high molecular weight proteins or protein aggregates. Since agarose gels do not contain catalysts which can influence the buffer system, they have also been used to develop a series of multiphasic discontinuous buffer systems (Jovin et al. 1970).

Jovin TM, Dante ML, Chrambach A. Multiphasic buffer systems output. Natl Techn Inf Serv. Springfield VA USA PB(1970)196 085-196 091.

Nucleic acids

Agarose electrophoresis is the standard method for separation, DNA restriction fragment-analysis and purification of DNA and RNA fragments (Maniatis et al. 1982; Rickwood and Hames, 1982). The fragment sizes analysed are in the range between 1,000 and 23,000 bp. Horizontal "submarine" gels are used for these nucleic acid separations: the agarose gel lies directly in the buffer (Fig. 12). This prevents the gel from drying out.

Maniatis T, Fritsch EF, Sambrook J. Molecular cloning a laboratory manual. Cold Spring Laboratory (1982).

Rickwood D, Hames BD. Gel electrophoresis of nucleic acids. IRL Press Ltd. (1982).

The gels are stained with fluorescent dyes like Ethidium bromide or SYBR Green, and the bands are visible under UV light. Their sensitivites ranges are between 100pg and 1 ng / band. Because they are intercalating in the helix, the sensitivity is dependent on the size of the DNA fragment and is lower for RNA detection.

These dyes have to be handled with care, because they are mutagens.

For a permanent record, mostly an instant photo is taken from the gels in a darkroom. Newly developed video documentation systems take the images inside a box print the results on thermopaper or feed them to a computer.

For RFLP (restriction fragment length polymorphism) analysis, the separated DNA fragments are transferred onto an immobilzing membrane followed by hybridization with radiolabelled probes (s. 4 Blotting).

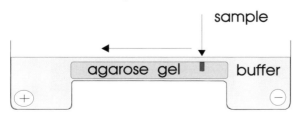

Fig. 12: The "submarine" technique for the separation of nucleic acids.

Pulsed field gel electrophoresis

For chromosome separation, pulsed field electrophoresis (PFG) according to Schwartz and Cantor (1984) is used; it is a modified submarine technique.

Schwartz DC, Cantor CR. Cell. 37 (1984) 67-75.

High molecular weight DNA molecules over 20 kb align themselves lengthwise during conventional electrophoresis and migrate with the same mobility so that no separation is achieved.

kb kilobases

In PFG the molecules must change their orientation with changes in the electric field, their helical structure is first stretched and then compressed. The *"viscoelastic relaxation time"* is dependent on the molecular weight. In addition, small molecules need more time to reorient themselves than large ones. This means that after renewed stretching and reorientation, larger molecules have - for a defined pulse - less time left for actual electrophoretic migration. The resulting electrophoretic mobility thus depends on the pulse time or on the duration of the electric field: a separation according to the molecular weight up to the magnitude of 10 megabases is obtained.

For shorter DNA fragments the resolution with PFG is also better than with conventional submarine electrophoresis.

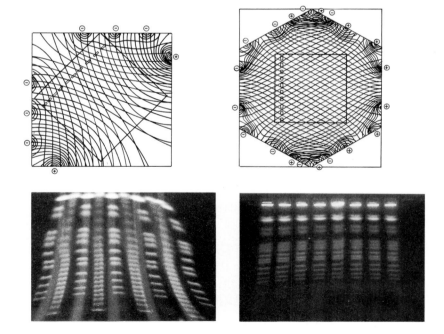

Fig. 13: Field lines and separation results for two types of PFG electrophoresis: *left* orthogonal doubly inhomogeneous fields and *right* homogeneous fields for hexagonally arranged point electrodes.

For the analysis of chromosomes, the sample preparation including cell disintegration, is done in agarose blocks which are placed in the pre-formed sample pockets. These molecules would be broken by the shear forces. 1.0 to 1.5% agarose gels are used for the separation.

The electric fields should have an angle of at least 110° relative to the sample. This is obtained for example by an inhomogeneous field with point electrodes mounted on orthogonal rails or in hexagonal configuration. The pulse time is of 1 s to 90 min for these techniques, depending on the length of the DNA molecules to be separated. Large molecules are better separated when the pulse time is long, small molecules needs short pulse times. The separations can last several days.

Fig. 13 shows the field lines for an orthogonal configuration with an inhomogeneous field and for an hexagonal configuration with a homogeneous field as well as the corresponding separations.

Pulsed field gel electrophoresis is mainly employed for research, but it has also found its place in routine analysis for bacterial taxonomy.

There are in addition other field geometries:
● *Field Inversion (FI) electrophoresis: the electric field is pulsed back and forth in one direction.*
● *Transverse Alternating Field electrophoresis (TAFE): The gel is mounted vertically in an aquarium-like tank and the field is pulsed back and forth between electrode pairs mounted on the top and the bottom of both sides of the gel.*

1.2.3 Polyacrylamide gel electrophoresis of nucleic acids

DNA sequencing

In the DNA sequencing methods according to Sanger and Coulson (1975) or Maxam and Gilbert (1977), the last step is electrophoresis in a polyacrylamide gel under denaturing conditions. The four reactions — containing variously long fragments of the DNA strand to be analyzed, each terminating with a specific base — are separated one beside the other. Determination of the order of the bands in these four lanes from the bottom to the top of the gel yields the base sequence, that is, the genetic information.

Tris-borate EDTA (TBE) buffer is used. To completely denature the molecule, the process is usually carried out at a temperature over 50 °C and in the presence of urea. Irregular heat distribution results in the *"smiling"* effect, when the bands are turned up at the ends. For this reason, it has proved effective to prewarm the gels with thermoplates independent from the electric field.

Manual sequencing: In the *manual* technique the bands are mostly revealed by autoradiography. Nucleotides or primers labelled with P^{32} or S^{35} are used. The gels are usually thinner than 0.4 mm since they must be dried for autoradiography.

Sanger F, Coulson AR. J Mol Biol. 94 (1975) 441-448.

Maxam AM, Gilbert W. Proc Natl AcadSci USA. 74 (1977) 560-564.

Smiling effect: When the temperature in the middle of the gel is higher than at the edges the DNA fragments migrate faster.

*In practice, **vertical** gel slabs are used, which are - in most cases - heated by the elctric field. An aluminum plate behind one of the glass plates distributes the heat evenly.*

Alternative nonradioactive detection methods have been developed:

- Chromogenic or chemiluminescent detection on a membrane after the separated DNA fragments have been tranferred from the gel.

  This requires biotinylated or fluorescent primers, nucleotides or probes.

- Silver staining of the gel.

  This requires cycle sequencing.

The use of wedge shaped gels has proved useful: they generate a field strength gradient which induces a compression of the band pattern in the low molecular weight area and enables the analysis of substantially more bases in one gel.

The samples are introduced in sample wells (formed in the gel by a sample comb during polymerization) with microcapillaries or syringes with an extra thin needle. A typical sequencing autoradiogram is shown in Fig. 14.

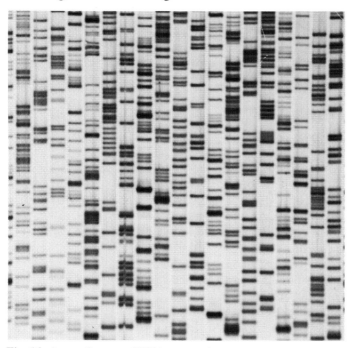

Fig. 14: Autoradiogram of DNA sequencing.

Automated sequencing: Samples with fluorescent tags are used. There are two principles:

Also here almost exclusively vertical slab gels are employed.

1. Single track system: For the four necessary reactions - with the base endings A, C, G, T - four different fluorescent markers are used. For separation, the four reagents are applied on the gel and the zones which migrate in one track are measured with selective photodetectors.

2. Four track system: This principle is based on the traditional Sanger method (Sanger and Coulson 1975). Only one dye is used, for example fluoresceine, which is used to mark the primer. The samples are separated in four tracks per clone. A fixed laser beam constantly scans the whole width of the gel in the lower fifth of the separation distance. At this height, a photovoltaic cell is fixed to the glass plate behind each band. When the migrating bands reach that spot, the fluorescent DNA fragments will be excited and emit a light signal (Ansorge *et al.* 1986). Since a single photo cell corresponds to each band, the migrating bands will be registered one after the other by the computer, giving the sequence. In one track systems, the raw data must be processed so that the mobility shifts due to the different markers are compensated. In four track systems, the sequence can be recognized directly from the raw data.

Ansorge W, Sproat BS, Stegemann J, Schwager C. J Biochem Biophys Methods. 13 (1986) 315-323.

Since the introduction of the Cy5 label, a red laser can be employed.

A

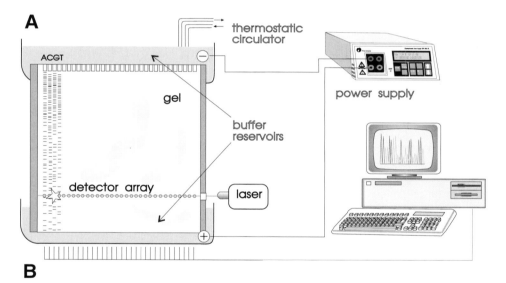

B

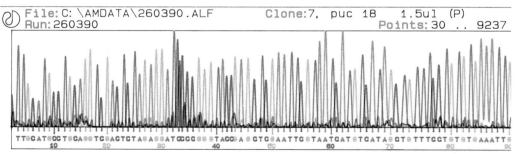

Fig 15: (A) Instrumentation for automated DNA sequencing with a four track system; (B) Typical trace after treatment of the crude data by a computer.

Automated sequencing has many advantages over the manual technique:

- Since fluorescent markers are used, the use of radioactivity in the laboratory can be avoided.

  No need for isotope laboratory.

- Neither extensive treatment of the gel after separation nor time-consuming autoradiography are necessary.

- The laborious reading of the bands becomes unnecessary.

- The sequences are directly fed into the computer.

- The reactions labelled with the fluorescent label can easily be kept for a long time, so that the separation can be repeated later in case of doubt.

- The high sensitivity of fluorescent labelling also allows the sequencing of cosmids and lambda DNA as well as the products of the polymerase chain reaction PCR®*). In addition restriction analyses can be carried out.

  *) The PCR process is covered by U.S. patents 4,683,195 and 4,683,302 owned by Hoffman-La Roche Inc. Use of the PCR process requires a license.

This "on-line electrophoresis" setup can also be employed for various DNA typing methods.

DNA typing

Many new techniques and applications have recently been developed in this field. Because those are almost exclusively based on PCR® technology, the size range of the DNA fragments to be analysed lies between 50 and 1,500 bp. In this range the sensitivity and resolution of agarose electrophoresis with Ethidium bromide staining is coming to its limits, because the gel pores are too large for proper sieving and the intercalating fluorescent dyes are much less sensitive than for larger fragments.

Amplification of fragments larger than 1,500 bp is possible, however, with a lot of problems with reproduciblity.

PAGE and silver staining:

The use polyacrylamide gels leads to much sharper bands and higher resolution; with subsequent silver staining a sensitivity of 15 pg per band can be achieved (Bassam *et al.* 1991). Vertical and horizontal slab gels can be used. Whereas in agarose electrophoresis the mobilities of DNA fragments are solely proportional to their sizes, the band positions in polyacrylamide gels are partly influenced by the base sequence as well. A and T rich fragments migrate slower than others.

Bassam BJ, Caetano-Annollés G, Gresshoff PM. Anal Biochem. 196 (1991) 80-83.

Silver staining of DNA is much easier than of proteins, because fixation is very easy.

Silver stained DNA bands can be directly reamplified after scratching them out of the gel without intermediate purification. About 20 % of the DNA molecules of a band remain undestroyed by the silver staining procedure. They are locked inside the stained band, thus DNA fragments do not contaminate the gel surface during staining.

Reamplification of DNA works only, when silver staining techniques specially designed for DNA detection are employed.

Horizontal electrophoresis

The horizontal system has a number of advantages over the vertical ones when ultrathin gels polymerized on support films are used (Görg *et al.* 1980): simpler handling, easy use of ready-made gels and buffer strips instead of large buffer volumes; good cooling efficiency; possibility of washing, drying and rehydrating the gels; automation.

Görg A, Postel W, Westermeier R, Gianazza E, Righetti PG, J Biochem Biophys Methods. 3 (1980) 273-284.

Random amplified polymorphic DNA (RAPD)

This method is applied for rapid detections of DNA polymorphisms of a wide variety of organisms: bacteria, fungi, plants, and animals. One single short oligonucleotide primer (10mer) of arbitrary sequence is used to amplify fragments of the genomic DNA (Welsh and McClelland, 1990; Williams *et al.* 1990). The low stringency annealing conditions lead to an amplification of a set of multiple DNA fragments of different sizes. When optimized and uniform PCR conditions are employed, specific and reproducible band patterns are achieved. A modification using 5mer primers is called DNA amplification fingerprinting (DAF) and has been introduced by Caetano-Annollés *et al.* (1991).

Welsh J, McClelland M. Nucleic Acids Res. 18 (1990) 7213-7218.

Williams JGK, Kubelik AR, Livak KJ, Rafalski JA, Tingey SV. Nucleic Acids Res. 18 (1990) 6531-6535.

Caetano-Annollés G, Bassam BJ, Gresshoff PM. Bio/Technology 9 (1991) 553-557.

RAPD samples can be run on agarose gels with Ethidium bromide staining or on polyacrylamide gels with subsequent silver staining. As the resolution and sensitivity of the latter method is much higher, more variety differences can be detected. Figure 16 shows the RAPD patterns of different fungus varieties separated in a horizontal polyacrylamide gel and silver stained.

Even one additional band detected can make a big difference in the evaluation.
With optimized separation and detection strain-specific pattern are achieved.
See Method 11.

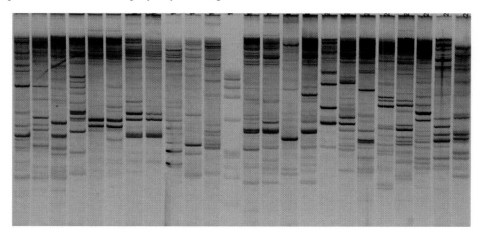

Fig. 16: RAPD electrophoresis of fungi varieties in a horizontal polyacrylamide gel. Silver staining. By kind permission of Birgit Jäger and Dr. Hans-Volker Tichy, TÜV Südwest GmbH - Biolocical Safety Division, Freiburg im Breisgau.

Amplified ribosomal DNA restriction analysis (ARDRA)

Fragments of ribosomal DNA with polymorphic restriction sites of an organism are amplified with a primer pair and subsequently digested with a restriction enzyme. After gel electrophoresis and silver staining, species specific pattern are obtained.

This method is derived from ribotyping and is mainly employed fro the identification of bacteria species.

Differential Display PCR Electrophoresis

This is a method to screen the total amount of cDNAs coming from the messenger RNA-pool of specific cell lines. The purpose is to display only the active genes besides the total amount of ca. 1 Million genes of a cell.

Not only cell regulation and differentiation can be monitored, but also miscontrolled cells can be visualized in cancer research.

The method "DDRT" (Differential display reverse transcription) has been introduced by Liang and Pardee (1992) and improved by Bauer *et al.* (1993). The mRNA from the original cell and the stimulated cell are processed in parallel. Extracted mRNA is reverse transcribed with oligo-dT-NN anchor primers. The resulting 12 cDNA pools are amplified with the respective oligo-dT primer and a set of arbitrary 10mer primers. After high resolution electrophoresis of the amplification products, those additional bands, which have been expressed by the stimulated cell, are cut out and reamplified for cloning and sequencing (see fig. 17). The original technique employes autoradiography for the detection of the bands.

Liang, P, Pardee AB. Science 257 (1992) 967-971.

Bauer D, Müller H, Reich J, Riedel H, Ahrenkiel V, Warthoe P, Strauss M. Nucleic Acid Res. 21 (1993) 4272-4280.

The additional bands are identified on the developed x-ray film; after cutting a hole in this position the film is matched with the gel again, the band is scratched out.

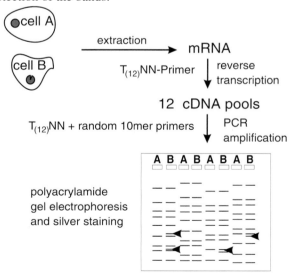

Additional bands are scraped out and reamplified for cloning and sequencing.

Fig. 17: The steps in a DDRT experiment.

Lohmann *et al* (1995) have taken a big step foreward with their "REN" (rapid, efficient, nonradioactive) technique: they use horizontal film-supported gels and cut out the silver stained DNA bands for reamplification. In this way, the method can be performed much faster, cheaper, and with a higher success of finding a gene, which has been expressed as a response of the cell.

Lohmann J, Schickle HP, Bosch TCG. BioTechniques 18 (1995) 200-202.
Urea and native gels can be employed. Sometimes it is necessary to use long gels, because the bands are spread over a wide range basepair-lengths.

Mutation detection methods

A comprehensive description of mutation detection methods can be found in the book "Laboratory Protocols for Mutation Detection", edited by Ulf Landegren (1996).

Laboratory Protocols for Mutation Detection. Landegren E, Ed. Oxford University Press (1996).

The most certain and sensitive method for the detection of mutations is the DNA sequence analysis. However, this method is too costly and time-consuming for screening purposes.

Single strand conformation polymorphism (SSCP)

The principle: Variations in the sequence as small as one base exchange alter the secondary structure of ssDNA, e.g. by different intramolecular base pairing. The changes in the sequence cause differences in the electrophoretic mobility, which are observed as band shifts (Orita *et al.* 1989).

Orita M, Iwahana H, Kanazewa H, Hayashi K, Sekiya T. Proc Natl Acad Sci USA. 86(1989) 2766-2770.

The mechanism of SSCP is described as: Differential transient interactions of the bent and curved molecules with the gel fibers during electrophoresis, causing the various sequence isomers to migrate with different mobilities.

Single strands migrate much slower than the corresponding double strands.

Before screening, the mutants have to be defined by direct sequencing. The sequences fo the appropriate primer pair have to be found. The PCR products are denatured by heating with formamide or sodium hydroxid, and loaded onto a non-denaturing polyacrylamide gel for electrophoresis. Silver staining has to be employed for detection of the DNA fragments. Many samples can be screened with a considerably lower effort than direct sequencing in a relatively short time, namely within a few hours.

SSCP analysis is not a replacement but an addition to sequencing, when 100 % of defined mutations have to be detected.

Intercalating dyes do not work here.

However, the band shifts do not show up automatically for all mutations and under all conditions. Unfortunately, there is not a single and unique separation condition, which can be applied to the separations of all exons. The parameters influencing the result have been reviewed by Hayashi and Yandell (1993) and will be further discussed in Method 12 in section II.

Hayashi K, Yandell DW. Hum Mutat. 2 (1993) 338 - 346.

For this method, good cooling and temperature control system is very important.

SSCP of the mitochondrial cytochrome b gene is also employed for differentiation of animal species. Rehbein *et al.* (1995) have used the method for the identification of the species in canned tuna.

Rehbein H, Mackie IM, Pryde S, Gonzales-Sotelo C, Perez-Martin R, Quintero J, Rey-Mendez M. Inf. Fischwirtsch. 42 (1995) 209-212.

Heteroduplex and DSCP

Single base substitutions can also be detected by heating the mixtures amplified wild type and mutant DNA and run the resulting heteroduplexes on a native polyacrylamide gel electrophoresis (Keen *et al.* 1991; White *et al.* 1992). The mobilities of heteroduplexes lie between the mobilities of the corresponding homoduplexes and single strands. Different mutations cause different mobility shifts of heteroduplexes. The bands can be detected with Ethidiumbromide or with silver staining.

Keen JD, Lester D, Inglehearn C, Curtis A, Bhattacharya. Trends Genet. 7 (1991) 5.

White MB, Carvalho M, Derse D, O'Brien SJ, Dean M. Genomics 12 (1992) 301-306.

Sometimes the technique is also called DSCP (double strand conformation polymorphism) (Barros *et al.* 1992). But it should not be forgotten, that also homoduplexes can show band shifts in native gels due to the influence of the contents of A and T.

Barros F, Carracedo A, Victoria ML, Rodriguez-Calvo MS. Electrophoresis 12 (1991) 1041-1045.

For DNA diagnosis, DNA point mutations can quickly be revealed with the *Primer Mismatch* process in combination with electrophoresis of the amplification products in horizontal polyacrylamide gels (Dockhorn-Dworniczak *et al.*, 1990).

Dockhorn-Dworniczak B, Aulekla-Acholz C, Dworniczak B. Pharmacia LKB Offprint A37 (1990).

Denaturing gradient gel electrophoreis (DGGE) and constant denaturing gel electrophoresis (CDGE)

With DGGE single base exchanges in segments of DNA can be detected with almost 100 % efficiency. The principle of DGGE is based on the different electrophoretic mobilities of partially denatured molecules caused by differences in DNA melting (Fischer and Lerman, 1983).

Fischer SG, Lerman LS. Proc Natl Acad Sci. 60 (1983) 1579-1583.

With a denaturant gradient perpendicular to the electrophoresis direction, the region of a point mutation can be identified. Denaturant gradients parallel to the electrophoresis runs are better for screening applications. Constant denaturing gel electrophoresis (CDGE) is employed for screening, when the denaturant concentration of differential melting of a DNA segment has been detected with DGGE. Figure 18 is a schematic representation of perpendicular and parallel DGGE. As DGGE is not very easy to perform, it is only employed, when the techniques other than sequence analysis fail in detecting a mutation.

Typically the 100 % denaturant solution contains 6 to 7 mol/L urea and 20 to 40 % formamide. The gels are run at temperatures between 40 ° and 60 °C.

The practical aspects and the gradient casting technique are described in Method 13.

Temperature gradient gel electrophoresis (TGGE)

Temperature gradient gel electrophoresis resolves homo- and heteroduplexes according to their thermal stabilities (Riesner *et al.* 1989). In this technique, denaturing gels are run on a plate with a cold (15 °C) side at the cathode and a hot side (60 °C) at the anode. The method is well suitable for screening purposes. Suttorp *et al.* (1996) have described how to change a standard horizontal electrophoretic chamber into a TGGE device.

Riesner D, Steger G, Wiese U, Wulfert M, Heibey M, Henco K. Electrophoresis 10 (1989) 377-389.
Suttorp M, von Neuhoff N, Tiemann M, Dreger P, Schaub J, Löffer H, Parwaresch R, Schmitz N. Electrophoresis 17 (1996) 672-677.

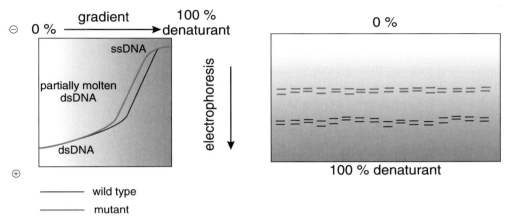

Fig. 18: Schematic representation of typical results of a perpendicular and a parallel DGGE.

Denaturing PAGE of microsatellites

Denaturing gels provide a very high resolving power, thus they are very useful for separating of microsatellites with very short repeats down to 2 bp. Because the Taq-polymerase used in PCR adds an additional A to the 3´-end of a part of the single strands, double bands are frequently seen after silver staining of denaturing gels. When labelling techniques like radioactivity or fluorescence are employed, only one of the pimer pair is marked to avoid visualization of the duplets.

It is not always necessary to apply completely denaturing conditons on the gel: 7 mol/L urea in the gel and 25° C separation temperature are often sufficient.

Instructions for denaturing electrophoresis are found in Method 14.

Native PAGE of mini- and microsatellites

Variable number of tandem repeats (VNTR) and short tandem repeats (STR) analysis are used in forensic laboratories: They are performed in denaturing and in non-denaturing gels. In both cases, the assignment of alleles with well-defined (sequenced) allelic ladders of the respective VNTR or STR locus, which are run in the same gel, proved to be the most reliable method (Puers *et al.* 1993; Möller *et al.* 1994). However, additionally to the regular types with length variations, there are sequence variants existing in some STR loci, which can only be identified by sequencing the fragments or running them on non-denaturing polyacrylamide gels.

Möller A, Wiegand P, Grüschow C, Seuchter SA, Baur MP, Brinkmann B. Int J Leg Med 106 (1994) 183-189.

Puers C, Hammond HA, Jin L, Caskey CT, Schumm JW. Am J Hum Genet 53 (1993) 953-958.

In order to achieve adequate resolution in native gels, long separation distances or special gel media and buffer systems have to be used (Schickle, 1996b).

Schickle HP. GIT Labormedizin. 19 (1996b) 228-231.

RNA and viroids

Bi-directional electrophoresis (Schumacher *et al*. 1986) is used for viroid tests: the plant extract (RNA fragment + viroid) is first separated under native conditions at 15 °C. After a certain separation time, the gel is cut behind a zone marked with a dye such as Bromophenol Blue or xylenecyanol.

Schumacher J, Meyer N, Riesner D, Weidemann HL. J Phytophathol. 115 (1986) 332-343.

An electrophoretic separation under denaturing conditions is carried out. The viroid forms a ring which cannot migrate. The RNA fragments which migrate more slowly during the first native separation, do not lose their mobility at 50 °C and migrate out of the gel. If a viroid is present, *only* one band is found when the gel is stained. The position of the viroid in the gel depends on its kind. Several new viroids have been discovered in this way.

The gel contains 4 mol/L urea. The molecules are denatured, that is unfolded by the combination of urea and elevated temperature. For practical reasons this method is only carried out in horizontal systems.

1.2.4 Polyacrylamide gel electrophoresis of proteins

For analytical PAGE of proteins, the trend is to go from cylindrical gels to flat and thinner ones. Because of the development of more sensitive staining methods such as silver staining for example, very small quantities of concentrated sample solution can be applied for the detection of trace amounts of proteins.
The advantages of thinner gels are:
- faster separation
- better defined bands
- faster staining
- better staining efficiency, higher sensitivity

Disc electrophoresis

Discontinuous electrophoresis according to Ornstein (1964) and Davis (1964) solves two problems during the separation of proteins in gels with small pores: it prevents the aggregation and precipitation of proteins during sample entry into the gel matrix and promotes the formation of well defined bands. The discontinuity is based on four parameters (see Fig. 19):
- the gel structure
- the pH value of the buffer
- the ionic strength of the buffer
- the nature of the ions in the gel and in the electrode buffer

Ornstein L. Ann NY Acad Sci. 121 (1964) 321-349.

Davis BJ. Ann NY Acad Sci. 121 (1964) 404-427.

The gel is divided into two areas: resolving and stacking gel. The resolving gel with small pores contains 0.375 mol/L Tris-HCL buffer pH 8.8, the stacking gel with large pores contains 0.125 mol/L Tris-HCL pH 6.8. In the electrode buffer terminating ions with low mobilities (e.g. glycine), in the gel exclusively leading ions with high mobility (e.g. Cl^-) are used as anions.

In the original work, chloride ions were used in the gel buffer. Some people use phosphate ions in gel.
Glycine is used as terminationg ion because it is very hydrophilic and does not bind to proteins.

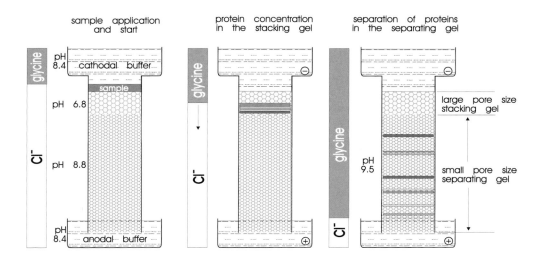

Fig. 19: Schematic diagram of the basic principles of disc electrophoresis according to Ornstein (1964).

Glycine has a very low net charge at the pH value of the stacking gel.

At first, the proteins are separated according to the principle of *isotachophoresis* and form stacks in the order of their mobility ("stacking effect"). The individual zones are concentrated. Because of the large pores in the gel matrix, the mobilities are dependent on the net charge, not on the size of the molecule.

The stacking effect is described in chapter 2, Isotachophoresis.

The protein stack migrates – slowly and at constant speed – towards the anode, till it reaches the limit of the separation gel. The frictional resistance suddenly increases for the proteins , they migrate slower, and the fractions become higher concentrated. The low molecular weight glycine is not affected by this, passes the proteins, and becomes more highly charged in the resolving zone; the new Cl^- / $glycine^-$ front moves ahead of the proteins.

Several events now occur simultaneously:

● The proteins are in a homogeneous buffer medium, destack and start to separate according to the principles of zone electrophoresis.

A discontinuity only exsist at the front.

● Their mobility now depends on their charge as well as on their size. The ranking of the protein ions changes.

● The pH value rises to 9.5 and because of this, the net charge of the proteins increases.

pK value of the basic groups of glycine

Disc electrophoresis affords high resolution and good band definition. In the example cited above, proteins with pIs higher than pH 6.8 migrate in the direction of the cathode and are lost.

Maurer RH. Disk- Electrophore-se - Theorie und Praxis der dis-kontinuierlichen Polyacrylamid-Electrophorese. W de Gruyter, Berlin (1968).

Another buffer system must be chosen to separate these proteins. A selection can be found in the works of Maurer (1968) and Jovin (1970).
The stacking gels is only cast onto the resolving gel just before electrophoresis because, when the complete gel is left standing for a long time the ions diffuse towards one another.

Gradient gel electrophoresis

By continuously changing the acrylamide concentration in the polymerization solution, a pore gradient gel is obtained. It can be used to determine the molecular diameter of proteins in their native state (Rothe and Purkhanbaba, 1982).

Rothe GM, Purkhanbaba M. Electrophoresis. 3 (1982) 33-42.

When the acrylamide concentration and the degree of cross-lin-king are fixed high enough in the small pore area, the protein molecules reach a point where, because of their size, they are trapped in the tight gel matrix. Since the speed of migration of the individual protein molecules depends on their charge, the electrophoresis must be carried out long enough so that the molecule with the lowest net charge also reaches its end point.

The determination of molecular weights in this manner can be problematic, since different pro-teins have different tertiary structures. Structural proteins cannot be compared with globu-lar proteins.

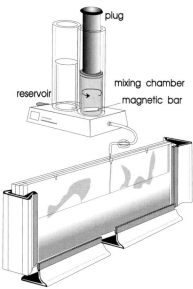

Fig 20: Casting of an exponential gradient gel with a gradient maker. The stirrer bar is rotated with a magnetic stirrer (not shown).

There are various ways of preparing gels with linear or exponential pore gradients. All are based on the same principle: two polymerization solutions with different acrylamide concentrations are prepared. During casting, the dilute solution is continuously mixed with the concentrated solution so that the concentration in the casting mold decreases from bottom to top (Fig. 20).

When single gradient gels are cast, the solution is poured into the top of the mould. When several gels are cast simultaneously, the solutions are injected from the bottom into multiple molds. In this case, the solutions in the mixing chamber and the reservoir are interchanged.

The density of the highly concentrated solution is increased with glycerol or sucrose so that the layers in the molds do not mix. In principle a concentration gradient is poured. The mixing of the less dense dilute solution with the viscous solution takes place in the mixing chamber using a magnetic stirrer bar.

If the mixing chamber is left open at the top, the principle of communicating vases is valid: so that the height of both fluids stays equal, half as much of the dilute solution flows in as of solution flowing out of the mixing chamber. A linear gradient is thus formed. A compensating stick in the reservoir compensates the volume of the stirrer bar and the difference in the densities of both solutions (see page 175 for a porosity gradient, pages 205 and 219 for pH gradients, and pages 209 and 249 ff for additive gradients).

Exponential gradients are formed when the mixing chamber is sealed. The volume in the mixing chamber stays constant, the same quantity of dilute solution flows in as solution out of the mixing chamber (see Fig. 20).

SDS electrophoresis

SDS electrophoresis − SDS being the abbreviation for *sodium dodecyl sulphate−* which was introduced by Shapiro *et al.* (1967) separates exclusively according to molecular weight. By loading with the anionic detergent SDS, the charge of the proteins is so well masked that anionic micelles with a constant net charge per mass unit result: 1.4 g SDS per g protein. In addition, the differences in molecular form are compensated by the loss of the tertiary and secondary structures because of the disruption of the hydrogen bonds and unfolding of the molecules.

Shapiro AL, Viñuela E, Maizel JV. Biochem Biophys Res Commun. 28 (1967) 815-822.

Disulfide bonds which can form between cysteine residues can only be cleaved by a reducing thiol agent such as 2-mercaptoethanol or dithiothreitol. The SH groups are often protected by a subsequent alkylation with iodoacetamide, iodoacetic acid or vinylpyridine (Lane, 1978).

Lane LC. Anal Biochem. 86 (1978) 655-664.

The unfolded amino acid chains, bound to SDS form ellipsoids with identical central axes. During electrophoresis in restrictive polyacrylamide gels containing 0.1% SDS there is a linear relationship between the logarithm of the molecular weight and the relative distance of migration of the SDS-polypeptide micelle.

This linear relationship is only valid for a certain interval which is determined by the ratio of the molecular size to the pore diameter.

Gels with a pore gradient have a wider separation interval and a larger linear separation interval than gels with a constant pore size. In addition, sharper bands result since a gradient gel minimizes diffusion (Fig. 21). The molecular weight of the proteins can be estimated with a calibration curve using marker proteins (Fig. 22).

Marker proteins for various molecular weight intervals exist.

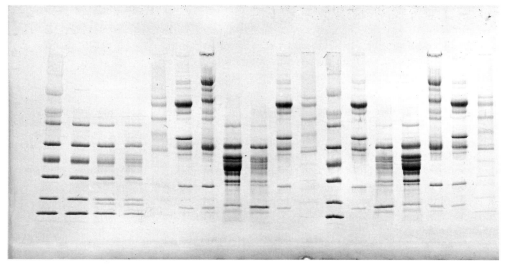

Fig. 21: Separation of proteins in a linear pore gradient gel T = 8% to 18% by SDS electrophoresis. Staining with Coomassie Brilliant Blue (Cathode on top).

For separation of physiological fluids or analysis of urine proteins for example, the reduction step is left out to prevent the breakdown of the immunoglobulins into subunits. In these cases the incomplete unfolding of certain proteins must be taken into account and therefore the molecular weight cannot be determined exactly.

For example when it is not reduced, albumin shows a molecular weight of 54 kDa instead of 68 kDa since the polypeptide chain is only partially unfolded.

There are a number of practical advantages to SDS electrophoresis:

- SDS solubilizes almost all proteins.

even very hydrophobic and denatured proteins

- Since SDS-protein complexes are highly charged, they possess a high electrophoretic mobility.

this ensures rapid separations

- Since the fractions are uniformly negatively charged, they all migrate in one direction.

towards the anode

- The polypeptides are unfolded and stretched by the treatment with SDS and the separation is carried out in strongly restrictive gels.

this limits diffusion

- This affords high resolution.

  sharp zones

- The bands are easy to fix.

  no strong acids are necessary

- The separation is based on one physico-chemical parameter, the molecular weight.

  It is an easy method for molecular weight determination.

- Charge microheterogeneities of isoenzymes are cancelled out.

  *There is **one** band for **one** enzyme.*

- Proteins separated with SDS bind dyes better.

  the detection limit increases tenfold compared to native PAGE

- After electrophoretic transfer on an immobilizing membrane, the SDS can be removed from the proteins without eluting the proteins themselves.

  see chapter 4: Blotting

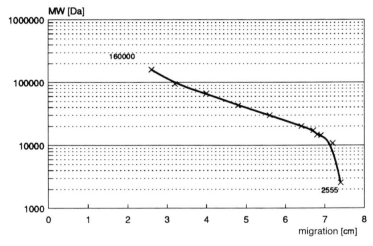

Fig. 22: Semi-logarithmic representation of a molecular weight curve. The molecular weights of the marker proteins are represented as a function of their migration. (SDS linear pore gradient gel according to Fig. 21)

SDS electrophoresis can be carried out in a continuous phosphate buffer system (Weber and Osborn, 1968) or in a discontinuous system:

Weber K, Osborn M: J Biol Chem. 244 (1968) 4406-4412.

Lämmli (1970) has directly taken over the disc electrophoresis method according to Ornstein (1964) and Davis (1964), for proteins charged with SDS, though the discontinuities in pH value and ionic strength are in most cases not necessary.

Lämmli UK. Nature. 227 (1970) 680-685.

- Because the protein-SDS micelles have very high negative charges, the mobility of glycine is lower than that of the proteins in the stacking gel at the beginning of electrophoresis, even at pH 8.8; it does not bind SDS.

  Glycine is not needed for the anodal electrophoresis buffer.

  The discontinuity of the anions is very important and the various gel porosities are very useful.

- During stacking no field strength gradient results, since there are no charge differences within the sample: so no low ionic strength is necessary.

This means that SDS disc electrophoresis gels can be cast in one step: Glycerol is added to the resolving gel and then the stacking gel, which contains the same buffer but no glycerol, is directly cast over it. In addition, the run time is shorter since the separation starts more quickly.

The overlayering of the resolving gel with butanol for example can thus be avoided and especially the laborious removal of the overlay before pouring the stacking gel.

Since there are no diffusion problems between the stacking and the resolving gel buffers with these gels, they can be stored longer than conventional disc gels. Yet their shelflife is limited by the high pH value of the gel buffer, since, after about 10 days, the polyacrylamide matrix starts to hydrolyse.

For ready-made gels with higher storage capacity, another buffer system with pH values around 7 should be chosen.

After empirical assays, a Tris-acetate buffer with a pH of 6.7 has proven to have the best storage stability and separation capacity. Tricine is used instead of glycine as the terminating ion. The principle of this buffer system with polyacrylamide electrode buffer strips in a ready-made SDS gel can be seen in Fig. 23.

Since Tricine is much more expensive than glycine, it is only used at the cathode, the anode contains Tris-acetate.

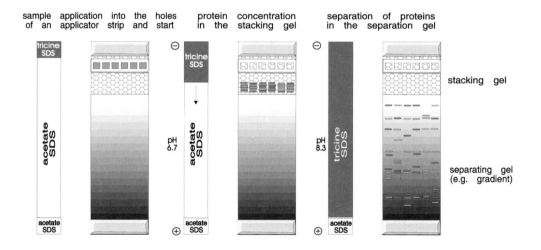

Fig. 23: Principle of the buffer systems of ready-made gels for discontinuous SDS-electrophoresis. Horizontal gels with Tris-Tricine buffer strips.

SDS-electrophoresis of low molecular weight peptides: the resolution of peptides below 14 kDa is not sufficient in conventional Tris-glycine-HCl systems. This problem has been solved by the development of a new gel and buffer system by Schägger and von Jagow (1987). In this method an additional spacer gel is introduced, the molarity of the buffer is increased and tricine used as terminating ion instead of glycine. This method yields a linear resolution from 100 to 1 kDa.

Schägger H, von Jagow G. Anal Biochem. 166 (1987) 368-379.

Glycoproteins: Glycoproteins migrate too slowly in SDS electrophoresis, since the sugar moiety does not bind SDS. When a Tris-borate-EDTA buffer is used , the sugar moieties are also negatively charged, so the speed of migration increases (Poduslo, 1981).

Poduslo JF. Anal Biochem. 114 (1981) 131-139.

The use of gradient gels is also beneficial for better MW estimations.

Membrane proteins: When membrane proteins are solubilized with nonionic detergents, these detergents would interfere with the SDS. Schägger and von Jagow (1991) have developed "Blue Native electrophoresis" to solve this problem:

Schägger H, von Jagow G. Anal Biochem. 199 (1991) 223-231.

In a vertical chamber Coomassie Blue G-250 is added to the cathodal buffer in a native polyacrylamide gel electrophoresis. During the run the dye competes with the nonionic detergent and binds to the membrane proteins (and some others) and charges the proteins negatively analogous to SDS. All these protein-dye complexes migrate towards the cathode, also basic proteins. They are soluble in detergent-free solution, and − as the negatively charged protein surfaces repel each other − aggregation between proteins is minimized.

The membrane proteins can be isolated in enzymatically active form.

The gels do not need to be stained, because the proteins migrate as blue bands.

Cationic detergents: Strongly acidic proteins do not bind SDS and very basic nucleoproteins behave abnormally in SDS gels. The alternative is to use a cationic detergent, cetyltrimethylammonium bromide (CTAB) in an acidic medium at pH 3 to 5 is recommended (Eley *et al.* 1979). This also allows a separation according to the molecular weight in the direction of the cathode. This cationic detergent also causes less damage to the protein than SDS, so CTAB electrophoresis can be used as a form of native electrophoresis (Atin *et al.* 1985).

Eley MH, Burns PC, Kannapell CC, Campbell PS. Anal Biochem. 92 (1979) 411-419.

Atin DT, Shapira R, Kinkade JM. Anal Biochem. 145 (1985) 170-176.

SDS electrophoresis in *washed, dried and rehydrated* polyacrylamide gels: the Tris-HCl / Tris-glycine buffer system shows very poor results. However good results are obtained with the Tris-acetate / Tris-tricine system. In this method, the gel is rehydrated in Tris-acetate pH 8.0 using a horizontal tray. If, for highly concentrated Protein samples, a discontinuity in pH and molarity between stacking and resolving gel is required, the stacking zone can be selectively equilibrated in a higher diluted Tris-acetate buffer pH 5.6 using a vertical chamber (see page 187).

The performance of SDS buffer systems are obviously highly influenced by catalysts and / or monomers of acrylamide.

This procedure of washing, drying, rehydration and equilibration can only be performed with gels polymerized on carrier films, which are used in horizontal systems.

Native electrophoresis in amphoteric buffers: The polymerization catalysts can be washed out of the polyacrylamide gels on support films used in horizontal systems with deionized water. By equilibration with amphoteric buffers such as HEPES, MES or MOPS for example, there is a wide spectrum for electrophoresis under native conditions.

The ionic catalysts APS and TEMED would destabilize these buffer systems, see method 4, pages 133 and following.

This method proved to be particularly useful for acidic electrophoresis of basic hydrophobic barley hordeins (Hsam et al. 1993) and basic fish sarcoplasmic proteins (Rehbein, 1995).

Hsam SLK, Schickle HP, Westermeier R, Zeller FJ.Brauwissenschaft 3 (1993) 86-94.

Rehbein H. Electrophoresis. 16 (1995) 820-822.

Two-dimensional electrophoresis

Several aims are pursued by the combination of two different electrophoretic methods.

- Proteins separated by electrophoresis are then identified by affinity- or immunoelectrophoresis according to Laurell (1966), or else more precisely characterized or quantified by crossed immunoelectrophoresis, for example.

- Part of the fractions of a complex protein mixture are separated by one method and then further purified by a second method based on other separation principles. Thus, irrelevant proteins do not interfere with the actual separation desired (Altland and Hackler, 1984).

- Mixtures of proteins are separated so that the physico-chemical parameters, such as pI and the molecular weight, can be read on the 2D electropherogram as on a coordinate system.

- Complex mixtures of proteins such as cell lysates or tissue extracts should be completely fractionated into individual proteins so as to obtain an overall picture of the protein composition and to enable location of individual proteins.

For these techniques, the first-dimensional runs are carried out in gel rods or strips and loaded onto the second-dimensional gels. A flat-bed gel can also be cut into strips after the first separation and transferred onto the second gel. In most cases the first-dimensional gel must be equilibrated in the buffer for the second-dimensional run to recharge the protein, for example to unfold the secondary and tertiary structure and to form a micelle with *SDS*.

High resolution 2D electrophoresis: During *"High Resolution 2D electrophoresis"* according to O'Farrell (1975) and Klose (1975), now a "classic" electrophoretic method, the first dimension is isoelectric focusing in presence of 8 or 9 molar urea – close to the saturation limit – and a non-ionic detergent such as Triton X-100 or Nonidet NP-40. The second dimension is an SDS electrophoresis. The separation parameter of the first-dimensional run, the pI, is independent of the molecular weight which is the separation parameter of the second-dimensional run In fig. 24 the principle of the classical 2D methodology is shown.

Abbreviation: 2D electrophoresis

Precipitin arcs which can be assigned to the zone of the first-dimensional run are thus obtained.

Altland K, Hackler R. In: Neuhoff V, Ed. Electrophoresis' 84. Verlag Chemie, Weinheim (1984) 362-378.

Such a map can be used as a layout for a protein data base.

O'Farrell PH. J Biol Chem. 250 (1975) 4007-4021.

Klose J. Humangenetik. 26 (1975) 231-243.

The denaturing conditions for the first dimension are necessary to prevent intermolecular interactions, keep hydrophobic proteins in solution, and avoid different conformations of one protein.

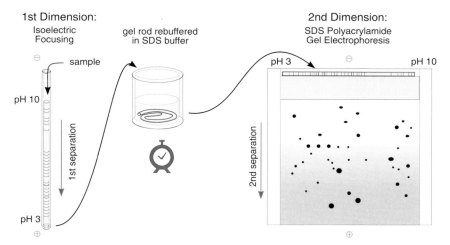

Fig. 24: The principle of the classical high-resolution 2D electrophoresis according to O`Farrell (1975) and Klose (1975).

The result of the separation is a pattern of spots. According to the Cartesian coordinate system, the following standard of representation has gained acceptance: from left to right — increasing pI, from bottom to top - increasing molecular weight. These two-dimensional protein maps afford the highest resolution of all the methods known at the moment. By lengthening the distance of separation, the use of thinner gels and the development of more sensitive methods of detection many research laboratories try to increase the number of detectable proteins as much as possible. Autoradiography of labelled proteins and silver staining are used most often. High reproducibility of the position of the spots is very important for interpretation.

The *IsoDalt* system, introduced by Anderson and Anderson (1978), is a completely developed methodical system and equipment for running many of these separations in parallel under identical conditions.

Anderson NG, Anderson NL. Anal Biochem. 85 (1978) 331-340.

Many problems occur when the first-dimensional run is done in individual cylindrical gels according to the traditional methods. The pH gradient begins to drift because of electro-endosmosis due to negative charges on the glass walls, which is particularly problematic because of the long separation times. This results in a variation of the spots positions; some of the proteins are lost, especially the basic ones.

The sample preparation and the type of sample application also have a noticeable influence and are directly linked with the technique of the first dimension.

In a practically oriented review article Dunn and Burghes (1983a) have listed a large number of physical and chemical parameters which influence the separation.

Dunn MJ, Burghes AHM. Electrophoresis. 4 (1983a) 97-116.

IPG-Dalt

The use of immobilized pH gradients (IPG) for the first-dimensional run allows the reproducibility of the pattern to be considerably increased independently of the separation time and buffer, and thus also allows the basic proteins to be detected (Görg *et al.* 1988a; Hanash and Strahler, 1989; Görg, 1991; Görg, 1993). It has been called the "IPG-Dalt" method following the name "Iso-Dalt" (Anderson and Anderson, 1978)

Even extremely wide immobilized pH gradients, e.g. pH 2.5 to 11, can be used, in order to separate nearly all possible cellular products in a single two-dimensional map (Sinha et al. 1992).

Practical details on this technique are found under method 10 in the second part of this book , and elsewhere (Görg, 1994).

Fig. 25 shows a 2D electropherogram of basic yeast cell proteins, which was run according to the IPG-Dalt system (Görg *et al.* 1988b) and silver stained according to Merrill *et al.* (1981). This separation is only possible when an immobilized pH gradient is used for the first-dimensional run.

Görg A, Postel W, Günther S. Electrophoresis. 9 (1988a) 531-546.

Hanash SM, Strahler JR. Nature. 337 (1989) 485-486.

Görg A. Nature . 349 (1991) 545-546.

Görg A. Biochem Soc Trans. 21 (1993) 130-132.

Sinha P, Köttgen E, Westermeier R, Righetti PG. Electrophoresis. 13 (1992) 210-214.

Görg A. In: Celis J, Ed. Cell Biology: A Laboratory Handbook. Academic Press Inc.San Diego, CA. (1994) 231-242.

Görg A, Postel W, Weser J, Günther S, Strahler JR, Hanash SM, Somerlot L, Kuick R. Electrophoresis. 9 (1988b) 37-46.

Merill CM, Goldman D, Sedman SA, Ebert MH. Science. 211 (1981) 1437-1438.

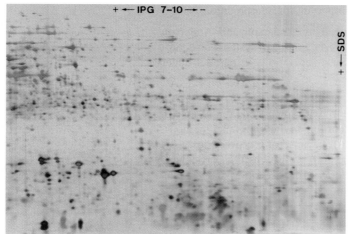

+ ← IPG 7-10 → –

SDS

Fig 25: High-resolution horizontal 2D electrophoresis of basic proteins from yeast cell lysate (Saccharomyces cerevisiae).
Immobilized pH-gradient in the first-dimension run (IPG-Dalt).
2nd dimension is a horizontal SDS porosity gradient gel. Silver staining according to Merill *et al.* (1981).
By kind permission of Professor A. Görg (1988b).

Evaluation of 2D patterns is performed with a computer. The pherograms are therefore converted into digital signals with densitometers, high resolution desk top scanners or video cameras. With the appropriate evaluation software the data are processed for qualitative and quantitative analysis, which is performed internally or with the help of federated 2D electrophoresis databases via international networks.

More detailed informations on this subject can be found in the second review article by Dunn and Burghess (1983b). Figure 26 is a photography from a computer screen, which displays the digitalized and processed image of several 2D electrophoresis gels from barley seed proteins. The histograms of spot intensities represent the quantitative contents of selected proteins in different varieties.

Dunn MJ, Burghes AHM. Electrophoresis. 4 (1983b) 173-189.

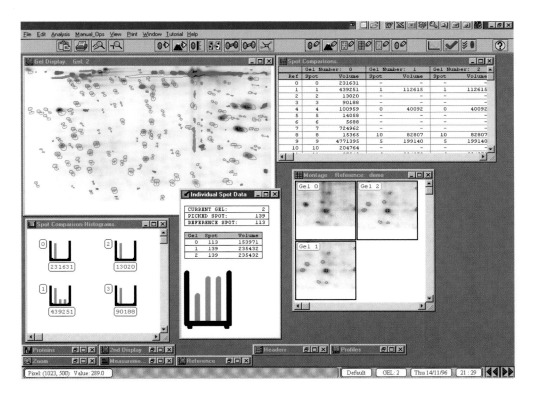

Fig. 26: Computer screen, which displays the digitalized image of several 2D electrophoresis gels.

2 Isotachophoresis

The main prerequisite for an isotachophoretic separation is a discontinuous buffer system with a leading and a terminating electrolyte. If the anions in a sample are to be determined, for example, the leading electrolyte should contain anions with higher mobility and the terminating electrolyte anions with a lower mobility than those of the sample which will be analyzed. In such an anionic separation, the leading electrolyte will be at the anodal and the terminating electrolyte at the cathodal side. The sample is applied between the two. The system also contains a common cationic counter-ion.

Either anions or cations can be separated at one time but not both simultaneously.

When an electric field is applied a potential gradient is induced between the two electrodes. So that all the ions migrate at the same velocity the field strength is higher in the vicinity of the ions of lower mobility than in the area containing the more mobile ions. The ions must migrate at the same speed, none can move too fast or drop behind (travel too slowly): otherwise no current could be carried.

See also Fig. 1.

During this migration pure contiguous zones containing the individual substances are formed within the sample. At equilibrium, ions with the highest mobility migrate in front of the terminating ion, the others migrate between them order of decreasing mobility:

The sample ions form stacks.

$$m_{L^-} > m_{A^-} > m_{B^-} > m_{T^-}.$$

m: mobility,
L^- leading ion,
T^- terminating ion,
A^- and B^- sample ions

The zone with the highest mobility has the lowest field strength, the one with the lowest mobility has the highest field strength. The product of the field strength and the mobility of each zone is constant. This has a *zone sharpening* effect: the ions which diffuse in a zone with a higher mobility will be slowed down because of the lower electric field and will migrate back to their own zone. Should an ion fall behind, it will be accelerated out of the neighbouring zone by the higher field.

The system works against diffusion and results in a distinct separation of the individual substances. In contrast to other separation techniques, the fractions are contiguous.

Isotachophoresis is carried out at constant current so as to maintain a constant field strength within the zones. The speed of migration then also stays the same during the separation.

Quantitative analysis

The basis of quantitative analysis with isotachophoresis is the "regulating function" (beharrliche Funktion) of Kohlrausch (1897). It defines the conditions at the boundary between two different ions L− and A− with the same counter-ion R+ during the migration of this boundary in the electrical field. The ratio of the concentrations C_L- and C_A- of the ions L^-, A^- and R^+ is the following:

Kohlrausch F. Ann Phys. 62
(1897) 209-220.

$$\frac{C_{L^-}}{C_{A^-}} = \frac{m_{L^-}}{m_{L^-} + m_{R^+}} \times \frac{m_{A^-} + m_{R^+}}{m_{A^-}}$$

m, the mobility is expressed in $cm^2/V \times s$ and is constant for each ion under defined conditions.

At a predefined concentration of the leading electrolyte L^-, the concentration of A− is fixed since all the other parameters are constant. This can be applied to the next zone: since the concentration of A^- is defined, the concentration of B^- is determined, and so on...

This leads to the concentrating effect: the higher the concentration of the leading ion, the more concentrated the zones.

Kohlrausch's equation can be more simply expressed as :

$$C_{A^-} = C_{L^-} \times \text{constant}$$

At equilibrium the concentration of the sample ions CA- is proportional to the concentration of the leading ions CL-. This means that the ionic concentration is constant in each zone. The number of ions in each zone is proportional to the length of the zone. A characteristic of isotachophoresis is that the quantification of the individual separated components is done by measuring the length of the zone, independently of the concentration of the original sample applied. Fig. 27 shows how, during the isotachophoretic run, states a and c automatically determine state b.

The bands are not "peaks" (Gaussian distribution) as in conventional electrophoresis or chromatography but "spikes" (concentration dependent bands). For this reason standard interpretation programs cannot be used.

To determine the concentration of a substance at least two runs must be performed: first the unmodified sample is separated, and then during the second run, a known amount of pure substance is added. The original quantity of the substance to be analyzed can be deduced from the lengthening of the zone.

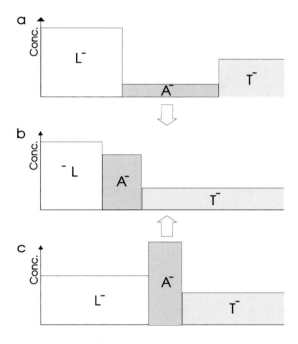

Fig. 27: The concentration regulating effect is the basis of quantification in isotachophoresis. Further explanations in the text.

Isotachophoresis is usually performed in Teflon capillaries (Everaerts *et al.* 1976; Hjalmarsson and Baldesten, 1981) new developments being quartz capillaries. Voltages up to 30 kV and currents of the order of several µA are used.
Teflon capillaries are used for detection of the zones by UV light after separation whereas values can be measured directly in quartz capillaries (Jorgenson and Lukacs, 1981; Hjertén, 1983). For effective differentiation between directly contiguous zones, current and thermometric conductivity detectors are also used.

Fig. 28 shows the isotachophoretic separation of penicillins: 10 µg each of ampicillin, phenoxymethylpenicillin and carbenicillin.

Everaerts FM, Becker JM, Verheggen TPEM. Isotachophoresis, Theory, instrumentation and applications. J Chromatogr Library. Vol. 6. Elsevier, Amsterdam (1976).

Hjalmarsson S-G, Baldesten A. In: CRC Critical Rev in Anal Chem. (1981) 261-352.

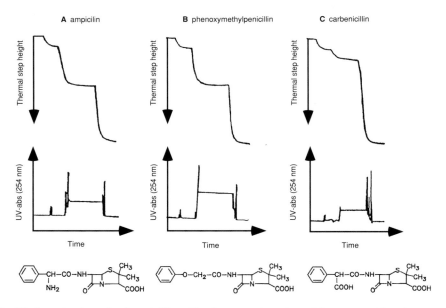

Fig. 28: Isotachophoresis of penicillins. Simultaneous detection of the zones with a thermocouple detector and an UV detector. From the application laboratory of Pharmacia LKB, Bromma, Sweden.

3 Isoelectric focusing

abbreviation: IEF

3.1 Principles

The use of isoelectric focusing is limited to molecules which can be either positively or negatively charged. Proteins, enzymes and peptides are such amphoteric molecules. The net charge of a protein is the sum of all negative and positive charges of the amino acid side chains, but the three-dimensional configuration of the protein also plays a role (Fig. 29).

The substances to be separated must have an isoelectric point at which they are not charged.

At low pH values, the carboxylic side groups of amino acids are neutral:

$$R\text{-}COO^- + H^+ \rightarrow R\text{-}COOH$$

At high pH values, they are negatively charged:

$$R\text{-}COOH + OH^- \rightarrow R\text{-}COO^- + H_2O$$

The amino, imidazole and guanidine side-chains of amino acids are positively charged at low pH values:

$$R\text{-}NH_2 + H^+ \rightarrow R\text{-}NH_3^+$$

at high pH values, they are neutral:

$$R\text{-}NH_3^+ + OH- \rightarrow R\text{-}NH_2 + H_2O$$

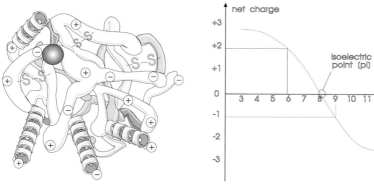

Fig. 29: Protein molecule and the dependence of the net charge on the pH value. A protein with this net charge has two positive charges at pH 6 and one negative charge at pH 9.

For composite proteins such as glyco- or nucleoproteins, the net charge is also influenced by the sugar or the nucleic acid moieties. The degree of phosphorylation also has an influence on the net charge.

Many of the microheterogeneities in IEF patterns are due to these modifications in the molecules.

If the net charge of a protein is plotted versus the pH (Fig. 29), a continuous curve which intersects the abscissa at the isoelectric point pI will result. The protein with the lowest known pI is the acidic glycoprotein of the chimpanzee: pI = 1.8. Lysozyme from the human placenta has the highest known pI: pI = 11.7.

The net charge curve is characteristic of a protein. With the titration curve method explained in chapter 3.8 it can easily be reproduced in a gel.

If a mixture of proteins is applied at a point in a pH gradient, the different proteins have a different net charge at this pH value (see Fig. 1).

It is important to find the optimum place in the gradient at which the proteins penetrate the gel without any trouble, do not aggregate and at which no protein is unstable.

In contrast to zone electrophoresis, isoelectric focusing is an end point method. This means, that the pattern – once the proteins have reached their pIs – is stable without time limit. Because of the focusing effect sharp protein zones and a high resolution are obtained.

The fact of the time stability of the pattern is not always true: carrier ampholytes pH gradients show a drift after some time, some proteins are not – or not very long – stable at their pI.

Isoelectric focusing is used with great success for protein isolation, on a preparative scale as well. It is also used for the identification of genetic variations and to investigate chemical, physical and biological influences on proteins, enzymes and hormones. In the beginning, sucrose concentration gradient columns in liquid phase were used, whereas gel media are almost exclusively employed nowadays.

The book by Righetti (1983) is recommended for further information.

The definition of the resolving power of isoelectric focusing was derived by Svensson (1961):

Svensson H. Acta Chem Scand. 15 (1961) 325-341.

$$\Delta pI = \sqrt{\frac{D[d(pH)/dx]}{E[-du/d(pH)]}}$$

ΔpI: resolution capacity
D: diffusion coefficient of the protein
E: field strength (V/cm)
$d(pH)/dx$: pH gradient
$du/d(pH)$: mobility slope at pI

ΔpI is the minimum pI difference needed to resolve two neighboring bands.
See also: titration curve analysis.

This equation shows how resolution can be increased:

● When the diffusion coefficient is high, a gel with small pores must be chosen so that diffusion is limited.

● A very flat pH gradient can be used.

But it also illustrates the limits of isoelectric focusing:

● Though the field strength can be raised by high voltages, it cannot be increased indefinitely.

● It is not possible to influence the mobility at the pI.

3.2 Gels for IEF

Analytical focusing is carried out in polyacrylamide or aga-rose gels. It is advantageous to use very thin gels with large pores cast onto films (Görg *et al.* 1978).

Görg A, Postel W, Westermeier R. Anal Biochem. 89 (1978) 60-70.

1. Polyacrylamide gels

Ready polymerized carrier ampholyte polyacrylamide gels and rehydratable polyacrylamide gels with or without immobi-lized pH gradients (explained below) are commercially available.

These ready-made gels are all polymerized on support films.

The use of washed, dried and rehydrated gels has been pub-lished soon after introduction of polyacrylamide for IEF byRo-binson (1972) , the methodology has been considerably impro-ved by Allen and Budowle (1986). The benefits are listed at page 151.

Robinson HK. Anal Biochem. 49 (1972) 353-366.
Allen RC, Budowle B, Lack PM, Graves G. In Dunn M, Ed. Elec-trophoresis '86. VCH, Weinheim (1986) 462-473.

Hydrophobic proteins need the presence of 8 to 9 molar urea to stay in solution. Because of the buffering capacity of urea, there is a light increase in the pH in the acid part of the gel. High urea contents in the gel lead to configurational changes in many proteins and disruption of the quaternary structure. The solubility of very hydrophobic proteins, such as membrane proteins for example, can be increased by the addition of non-ionic deter-gents (e.g. Nonidet NP-40, Triton X-100) or zwitterionic deter-gents (e.g. CHAPS, Zwittergent).

Because the gels do not co-poly-merize with the support films in the presence of non-ionic deter-gents, it is recommended to re-hydrate a prepolymerized, washed and dried gel in the rele-vant solution.

2. Agarose gels

Agarose gels for isoelectric focusing have only been available since 1975, when it became possible to eliminate the charges of agarose by removing or masking the agaropectin residues in the raw material. Agarose IEF exhibits stronger electroendosmosis than polyacrylamide gel electrophoresis IEF.

Carboxylic and sulfate groups which can be charged always remain.

Separations in agarose gels, usually containing 0.8 to 1.0% agarose, are more rapid. In addition macromolecules larger than 500 kDa can be separated since agarose pores are substantially larger than those of polyacrylamide gels. Agarose gels are also often used for IEF because its components – in contrast to those used for polyacrylamide gels – are not toxic and do not contain catalysts which could interfere with the separation.

Disadvantages:
silver staining does not work as well for agarose gels as for po-lyacrylamide gels. In the basic area, electroendosmosis is parti-cularly strong.

It is difficult to prepare stable agarose gels with high urea concentrations because urea disrupts the configuration of the helicoidal structure of the polysaccharide chains. Rehydratable agarose gels are advantageous in this case (Hoffman *et al.* 1989).

Hoffman WL, Jump AA, Kelly PJ, Elanogovan N. Electrophoresis. 10 (1989) 741-747.

3.3 Temperature

Since the pK values of the Immobilines, the carrier ampholytes and the substances to be analyzed are temperature dependent, IEF must be carried out at a constant controlled temperature, usually 10 °C. For the analysis of the configuration of subunits of specific proteins, ligand bindings or enzyme-substrate complexes, cryo-IEF methods at temperatures below 0 °C are used Righetti (1977). In order to increase the solubility of cryoproteins (like IgM), agarose IEF is performed at + 37 °C.

Righetti PG. J. Chromatogr. 138 (1977) 213-215.

Cryoproteins are precipitating at low temperatures

3.4 Controlling the pH gradient

Measurement of the pH gradient with electrodes is a problem since these react very slowly at low temperatures. In addition additives influence the measurement. CO_2 diffusing into the gel from the air reacts with water to form carbonate ions. Those form the anhydrid of carbonic acid and lowers the pH of the alkaline part. To prevent errors which can occur during the measurement of pH gradients, it is recommended to use marker proteins of known pIs. The pIs of the sample can then be measured with the help of a pH calibration curve.

Marker proteins for various pH ranges exist. These proteins are chosen so that they can focus independently of the point of application.

Note: Standard marker proteins can not be used in urea gels, because their conformations are changed, and thus their pIs.

3.5 The kinds of pH gradients

The prerequisite for highly resolved and reproducible separations is a stable and continuous pH gradient with regular and constant conductivity and buffer capacity.

There are two different concepts which meet these demands: pH gradients which are formed in the electric field by amphoteric buffers, the carrier ampholytes, or immobilized pH gradients in which the buffering groups are part of the gel medium.

3.5.1 Free carrier ampholytes

The theoretical basis for the realization of *"natural"* pH gradients was derived by Svensson (1961) while the practical realization is the work of Vesterberg (1969): the synthesis of a heterogeneous mixture of isomers of aliphatic oligoamino-oligocarboxylic acids. These buffers are a spectrum of low molecular weight ampholytes with closely related isoelectric points.

Vesterberg, O. Acta Chem. Scand. 23 (1969) 2653-2666.

The general chemical formula is the following:

$$- CH_2 - N - (CH_2)_x - N - CH_2 -$$

with $(CH_2)_x$ and NR_2 on the first nitrogen branch and $(CH_2)_x$ and $COOH$ on the second:

$$
\begin{array}{cc}
(CH_2)_x & (CH_2)_x \\
| & | \\
NR_2 & COOH
\end{array}
$$

Where R = H or -(CH₂)ₓ-COOH, x = 2 or 3

These carrier ampholytes possess the following properties:
- a high buffering capacity and solubility at the pI,
- good and regular conductivity at the pI,
- absence of biological effects,
- a low molecular weight.

Naturally occurring ampholytes such as amino acids and peptides do not have their highest buffering capacity at their isoelectric point. They can therefore not be employed.

The pH gradient is produced by the electric field. For example, in a focusing gel with the usual concentration of 2 to 2.5% (w/v) carrier ampholyte (e.g. for gradients from pH 3 to 10) the gel has a uniform average pH value. Almost all the carrier ampholytes are charged: those with the higher pI positively, those with the lower pI negatively (Fig. 30).

By controlling the synthesis and the use of a suitable mixture the composition can be monitored so that regular and linear gradient result.

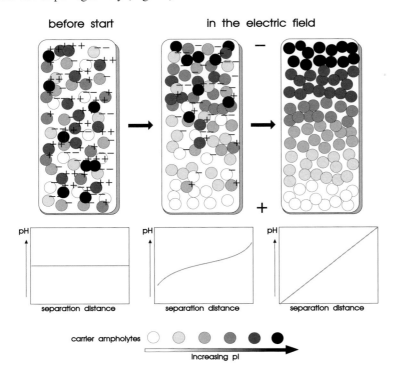

Fig. 30: Diagram of the formation of a carrier ampholyte pH gradient in the electric field.

When an electric field is applied, the negatively charged carrier ampholytes migrate towards the anode, the positively charged ones to the cathode and their velocity depends on the magnitude of their net charge.

The anodic end of the gel becomes more acidic and the cathodic side more basic.

The carrier ampholyte molecules with the lowest pI migrate towards the anode and those with the highest pI towards the cathode. The other carrier ampholytes align themselves in between according to their pI and will determine the pH of their environment. A stable, gradually increasing pH gradient from pH 3 to 10 results (Fig. 30).

The carrier ampholytes lose part of their charge so the conductivity of the gel decreases.

Since carrier ampholytes have low molecular weights they have a high rate of diffusion in the gel. This means that they diffuse away from their pI constantly and rapidly and migrate back to it electrophoretically: because of this, even when there are only a limited number of isomers a "smooth" pH gradient results. This is particularly important when very flat pH gradients, for example between pH 4.0 and 5.0, are used for high resolution.

The proteins are considerably larger than the carrier ampholytes - their diffusion coefficient is considerably smaller - they focus in sharper zones.

Electrode solutions

To maintain a gradient as stable as possible, strips of filter paper soaked in the electrode solutions are applied between the gel and the electrodes, an acid solution is used at the anode and a basic one at the cathode. Should, for example, an acid carrier ampholyte reach the anode, its basic moiety would acquire a positive charge from the medium and it would be attracted back by the cathode.

These electrode solutions are particularly important for long separations in gels containing urea, for basic and for flat gradients. They are not necessary for short gels.

The native IEF in Fig. 31 could be carried out without electrode solutions because a washed and rehydrated gel with a wide pH gradient was used.

Electrode solutions for agarose IEF are listed in chapter 5, for polyacrylamide gels in chapter 6 of sectorII.

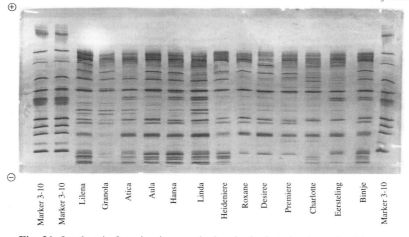

Fig. 31: Isoelectric focusing in a washed and rehydrated polyacrylamide gel. Press sap of potatoes of different varieties. Staining with Coomassie Brilliant Blue. (Anode on top). From Pharmacia EuroLab, Freiburg.

Separator IEF

Ever since the introduction of IEF, modifications of the pH gradients have been investigated. If the resolution is not satisfactory, it is often possible to add *separators* (Brown *et al*. 1977):

Brown RK, Caspers ML, Lull JM, Vinogradov SN, Felgenhauer K, Nekic M. J Chromatogr. 131 (1977) 223-232.

These are amino acids or amphoteric buffer substances which flatten the pH gradient in the area of their pI. Their position in the gradient can be changed by adapting the temperature conditions and separator concentration so that complete separation of neighbouring protein bands can be achieved.

One example is the separation of glycosylated HbA from the neighbouring main hemoglobin band in the pH gradient 6 to 8 by the addition of 0.33 mol/L ß-alanine at 15 °C (Jeppson et al. 1978).

Jeppson JO, Franzen B, Nilsson VO. Sci Tools. 25 (1978) 69-73.

Plateau phenomenon

Problems with carrier ampholytes can arise when long focusing times are necessary. For example, when the gradients are small or in the presence of highly viscous additives such as urea or non-ionic detergents, the gradient begins to drift in both directions but specially towards the cathode. This leads to a plateau in the middle with gaps in the conductivity. Part of the proteins migrate out of the gel (Righetti and Drysdale, 1973) and are not included. Because of the limited number of different homologues, the gradients cannot be flattened and the resolution capacity not increased at will.

Righetti PG, Drysdale JW. Ann NY Acad Sci. 209 (1973) 163-187.

A gel can "burn" through at the conductivity gaps.

The procedure of a carrier ampholyte IEF run

As isoelectric focusing is in principle a nondenaturing method, the optimization of the running conditions is very important to prevent precipitation and aggregation of proteins, and to achieve good reproducibility.

The IEF running conditions should always be given in a proteocol or a publication.

● Temperature setting

pIs are highly dependent on the temperature.

● Prefocusing

To establish the gradient.

● Sample loading

On the optimized location with the optimized mode.

● Sample entry

At low field strength to prevent aggregation.

● Separation time is a compromise between letting all proteins reach their pIs and keeping the gradient drift to a minimum.

Votlhour integration is often used.

● Fixing (with TCA or by immunofixation) and staining – or alternatively – application of zymogram detection

Proteins have to be fixed because the carrier ampholytes have to be washed out.

3.5.2 Immobilized pH gradients

Because of some limitations of the carrier ampholytes method, an alternative technique was developed: immobilized pH gradients or *IPG* (Bjellqvist *et al.* 1982). This gradient is built with acrylamide derivatives with buffering groups, the Immobilines, by co-polymerization of the acrylamide monomers in a polyacrylamide gel.

Bjellqvist B, Ek K, Righetti PG, Gianazza E, Görg A, Westermeier R, Postel W. J Biochem Biophys Methods. 6 (1982) 317-339.

The general structure is the following:

$$CH_2 = CH - \underset{\underset{O}{\|}}{C} - \underset{\underset{H}{|}}{N} - R$$

R contains either a carboxylic or an amino group.

An *Immobiline* is a weak acid or base defined by its pK value.

At the moment the commercially available ones are:

- two acids (carboxylic groups) with pK 3.6 and pK 4.6.
- four bases (tertiary amino groups) with pK 6.2, pK 7.0, pK 8.5 and pK 9.3.

To be able to buffer at a precise pH value, at least two different Immobilines are necessary, an acid and a base. Fig. 32 shows a diagram of a polyacrylamide gel with polymerized Immobilines, the pH value is set by the ratio of the Immobilines in the mixture.

The wider the pH gradient desired, the more Immobiline homologues are needed.

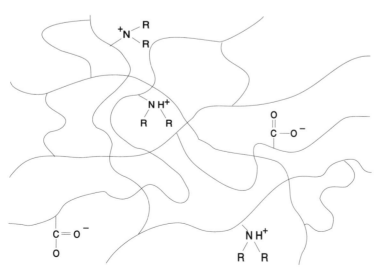

Fig. 32: Diagram of a polyacrylamide network with co-polymerized Immobilines.

A pH gradient is obtained by the continuous change in the ratio of Immobilines. The principle is that of an acid base titration and the pH value at each stage is defined by the *Henderson-Hasselbalch* equation:

Here the pH gradient is absolutely continuous.

$$pH = pK_B + \log \frac{C_B - C_A}{C_A}$$

C_A and C_B are the molar concentrations of the acid, and basic Immobiline, respectively.

when the buffering Immobiline is a base.

If the buffering Immobiline is an acid, the equation becomes:

$$pH = pK_A + \log \frac{C_B}{C_A - C_B}$$

Preparation of immobilized pH gradients

In practice immobilized pH gradients are prepared by linear mixing of two different polymerization solutions with a gradient maker (see Fig. 17), as for pore gradients. In principle a concentration gradient is poured. Both solutions contain acrylamide monomers and catalysts for the polymerization of the gel matrix.

0.5 mm thick Immobiline gels, polymerized on a support film have proved most convenient.

Immobiline stock solutions with concentrations of 0.2 mol/L are used. The solution which is made denser with glycerol is at the acid end of the desired pH gradient, the other solution is at the basic end. During polymerization, the buffering carboxylic and amino groups covalently bind to the gel matrix.

The catalysts must be washed out of the gel because they interfere with IEF. This is more rapid if the gel is thin.

Applications of immobilized pH gradients

Immobilized pH gradients can be exactly calculated in advance and adapted to the separation problem. Very high resolution can be achieved by the preparation of very flat gradients with up to 0.01 pH units per cm.

It has proved very practical to dry the gels after washing them and to let them soak in the additive solution afterwards.

Since the gradient is fixed in the gel it stays unchanged during the long separation times which are necessary for flat gradients, but also in IEF when viscous additives such as urea and non-ionic detergents are used. In addition there are no wavy iso-pH lines: the gradient is not influenced by proteins and salts in the solution.

Recipes for the preparation of narrow and wide immobilized pH gradients are given in this book in the section on methods for immobilized pH gradients (section II, method 9). The quantities necessary for the 0.2 molar Immobiline stock solutions for the acid and basic starter solutions are given in mL for the standard gel volume.

The broadest pH gradient which can, at present, be prepared with commercially available Immobilines encompasses 6 pH units: from 4.0 to 10.0.

Further developments

Altland (1990) and Giaffreda *et al.* (1993) have published programs for personal computers which permit the calculation the desired pH gradients with optimization of the distribution of buffer concentration and ionic strength.

Altland K. Electrophoresis. 11 (1990) 140-147.
Giaffreda E, Tonani C, Righetti PG. J Chromatogr. 630 (1993) 313-327.

In the meantime, thanks to the work of Righetti's group, it has been possible to expand the pH range mentioned before in both directions by using additional types of Immobilines and also to prepare very acidic (Chiari M *et al.* 1989a) and basic narrow pH gradients (Chiari M *et al.* 1989b). These are an additional acid with pK 0.8 and a base with pK 10.4. At these pH extremities the buffering capacity of the water ions H^+ and OH^- must be taken into consideration. Furthermore dramatic differences occur in the voltage gradient which must be compensated by the gradual addition of additives to the gradient.

Acidic Immobiline:
Chiari M, Casale E, Santaniello E, Righetti PG. Theor Applied Electr. 1 (1989a) 99-102.

Basic Immobiline:
Chiari M, Casale E, Santaniello E, Righetti PG. Theor Applied Electr. 1 (1989b) 103-107.

The use of immobilized pH gradients is at present restricted to polyacrylamide gels only.

This means that the pore size is limited towards the top.

Fig. 33 shows an IEF of α_1-antitrypsin isoforms in IPG pH 4.0 to 5.0.

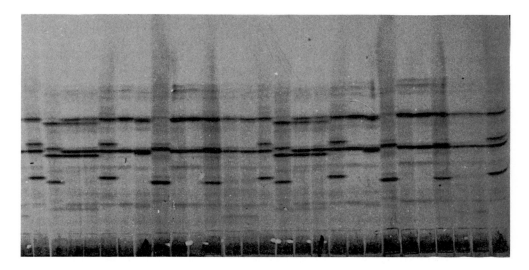

Fig. 33: IEF in immobilized pH gradient pH 4.0 to 5.0. Isoforms of α_1-antitrypsin (protease inhibitors) in human serum. By kind permission of Prof. Dr. Pollack and Ms. Pack, Institut für Rechtsmedizin der Universität Freiburg im Breisgau. (Anode at the top).

The technique of immobilized pH gradients offers so many new possibilities that it is constantly undergoing further developments. Many developments are presented in detail in the book specialized on IPGs by Righetti (1990).

Righetti PG.: Immobilized pH gradients: theory and methodology. Elsevier, Amsterdam (1990).

3.6 Preparative isoelectric focusing

1. Carrier ampholyte IEF

Preparative carrier ampholyte IEF is mainly carried out in horizontal troughs in granular gels (Radola, 1973). A highly purified dextran gel is mixed with the carrier ampholyte and poured in the trough. Here focusing is done over a long separation distance: about 25 cm. After prefocusing to establish the pH gradient a section of the gel is removed from a specific part of the gradient, mixed with the sample and poured back into place.

After IEF the protein or enzyme zones can be detected by staining a paper replica. To recover them, the gel is fractionated with a lattice and the fractions eluted out of the gel with a buffer. Proteins quantities of the order of 100 mg can thus be isolated.

Radola BJ. Biochim Biophys Acta. 295 (1973) 412-428. The procedure is described in: Westermeier R: In Doonan S. Ed. Protein Purification Protocols. Methods in Molecular Biology 59. Humana Press, Totowa, NJ (1996) 239-248.

For the elution, small columns with nylon sieves are used.

2. Immobilized pH gradients

Immobilized pH gradients are also very useful for preparative separations:

● They offer a high loading capacity.

● The buffering groups are fixed in the gel.

● The conductivity is low, so even gels which are 5 mm thick hardly heat up.

Polyacrylamide gels with IPG bind proteins more strongly than other media, so electrophoretic elution methods must be used (Righetti and Gelfi, 1984).

Righetti PG, Gelfi C. J Biochem Biophys Methods. 9 (1984) 103-119.

This technique is especially useful for low molecular peptides since the buffering groups of the gradient stay in the gel (Gianazza et al. 1983). Peptides are the same size and − after IEF − possess the same charge as the carrier ampholytes so they cannot be separated.

Gianazza E, Chillemi F, Duranti M, Righetti PG, J Biochem Biophys Methods. 8 (1983) 339-351.

Isoelectric membranes: An important approach is the application of the principles and chemistry of immobilized pH gradients on high resolution separation of proteins in a gel-free liquid. Righetti et al. (1989) have designed a multicompartment apparatus, whose segments are divided by isoelectric Immobiline membranes.

The electrodes are located in the two outer segments. The separation happens between the isoelectric membranes in gel-free liquid, which is constantly recirculated.

Righetti PG, Wenisch E, Faupel M. J Chromatogr. 475 (1989) 293-309.

The apparatus is now commercially available.

The highlight of the system are the membranes with defined pH values (Wenger *et al.* 1987): glass microfiber filters are soaked in acrylamide polymerization solutions, which are titrated exactly to the desired pH values with Immobilines. Thus "crystal grade" proteins are obtained without further contamination.

Fig. 34 shows the principle of the purification of a protein in a three-chamber setup, where the chambers are divided by two membranes with pH values closely below and above the pI of the protein to be purified.

Wenger P, de Zuanni M, Javet P, Righetti PG. J Biochem Biophys Methods 14 (1987) 29-43.

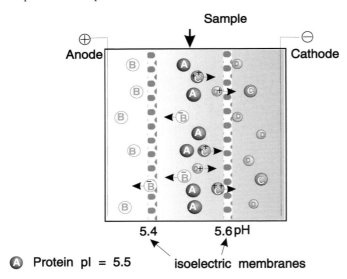

Fig. 34: Purification of a protein (A) between two isoelectric membranes. All contaminating charged substances and proteins with pIs lower than pH 5.4 (B) or higher than pH 5.6 (C,D) migrate out of the central chamber, which is enclosed by the two membranes.

3.7 Titration curve analysis

Carrier ampholyte gels can also be used to determine the charge intensity curve of proteins. This method is very useful for several reasons: it yields extensive information about the characteristics of a protein or enzyme, for example the increase in mobility around the pI, conformational changes or ligand bindings properties depending on the pH. The pH optimum for separation of proteins with ion-exchange chromatography and for preparative electrophoresis can also be established (Rosengren et al. 1977).

Rosengren A, Bjellqvist B, Gasparic V. In: Radola BJ, Graesslin D. Ed. Electrofocusing and isotachophoresis. W. de Gruyter, Berlin (1977) 165-171.

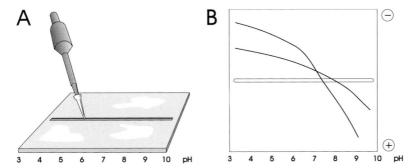

Fig. 35: Titration curves. A) Application of the sample in the sample trench *after* the pH gradient has been established. B) Titration curves.

A pre-run is performed in the square gel without any sample until the pH gradient is established. The gel, placed on the cooling plate, is rotated by 90° and the sample is applied in a long trough previously polymerized into the gel (see Fig. 35 A).

When an electric field is applied perpendicular to the pH gradient, the carrier ampholytes will stay in place since their net charges are zero at their pIs.

The sample proteins will migrate with different mobilities according to the pH value at each point and will form curves similar to the classical acid-base titration curves (Fig. 35 B). The pI of a protein is the point at which the curve intersects the sample trough.

There is a representation standard for titration curves for purposes of comparison: the gel is oriented so that the pH values increase from left to right and the cathode is on top (Fig. 36).

A gel with large pore sizes (4 to 5 % T) is used, in order to avoid influence s of the molecule sizes on the mobilities.

In practice a series of native electrophoresis runs under various pH conditions are carried out.

As can be seen, no buffer reservoirs are necessary for native electrophoresis in amphoteric buffers. This forms the basis of an electrophoretic method which is described in method 4.

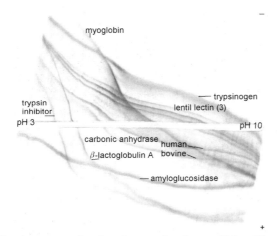

Fig. 36: Titration curves of a pI marker protein mixture pH 3 to 10. Cathode at the top.

4 Blotting

4.1 Principle

Blotting is the transfer of large molecules on to the surface of an immobilizing membrane. This method broadens the possibilities of detection for electrophoretically separated fractions because the molecules adsorbed on the membrane surface are freely available for macromolecular ligands, for example antigens, antibodies, lectins or nucleic acids. Before the specific detection the free binding sites must be blocked with substrates which do not take part in the ensuing reaction (Fig. 37).

In addition blotting is an intermediate step in protein sequencing and an elution method for subsequent analyses.

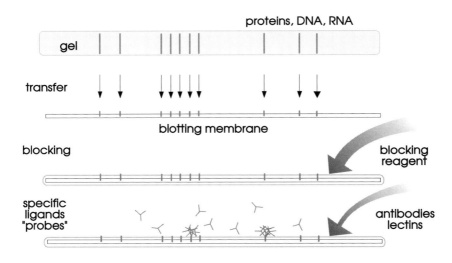

Fig. 37: The most important steps during blotting from electrophoresis gels.

4.2 Transfer methods

Diffusion blotting

The blotting membrane is applied on to the gel surface as when making a replica. The molecules are transferred by diffusion. Since the molecules diffuse regularly in every direction, the gel can be placed between two blotting membranes thus yielding two mirror-image transfers (Fig. 38). The diffusion can be accelerated by increasing the temperature, the technique in then known as *thermoblotting*. It is mostly used after electrophoresis in gels with large pores.

Quantitative transfers cannot be achieved with this method, especially not with larger molecules.

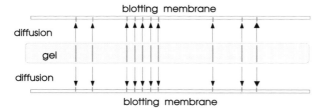

Fig. 38: Bi-directional transfer of proteins by diffusion blotting from a gel with large pores.

Capillary blotting

This technique is a standard one for subsequent hybridization according to Southern (1975) *(Southern blot)* during DNA separations. The transfer of RNA on to a covalently binding film or nylon membrane which is now known under the name *Northern blot* also uses this technique (Alwine *et al.* 1977).

Southern EM. J Mol Biol. 98 (1975) 503-517.
Alwine JC, Kemp DJ, Stark JR. Proc Natl Acad Sci USA, 74 (1977) 5350-5354.

This kind of transfer can also be used for proteins which were separated in a gel with large pores (Olsson *et al*, 1987). Buffer is drawn from a reservoir through the gel and the blotting membrane to a stack of dry paper tissues by capillary force. The molecules are carried to the blotting membrane on which they are adsorbed. The transfer occurs overnight (Fig. 39).

Olsson BG, Weström BR, Karlsson BW. Electrophoresis. 8 (1987) 377-464.

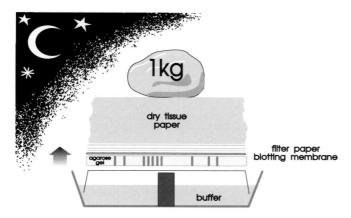

Fig. 39: Capillary blotting, the transfer occurs overnight.

Pressure blotting

Pressure blots from agarose gels cast on GelBond film are obtained very easily: a wet blotting membrane is laid on the gel, covered by one or several dry filter paper sheets, a glass plate and a 1 kg weight for 100 cm^2. The transfer is very fast, only a few seconds! Even multiple successive and identical blots can be obtained from one singele gel (Desvaux *et al.* 1990).

Desvaux FX, David B, Peltre G. Electrophoresis 11 (1990) 37-41.
This technique can be also employed for thin polyacrylamide IEF gels, however with a longer contact time (about 1 hour).

Vacuum blotting

This technique is mostly used instead of capillary blotting (Olszewska and Jones, 1988). It is important to have a controlled low vacuum with, depending on the case, a 20 to 40 cm high water column to prevent the gel matrix from collapsing. An adjustable pump is used since a water pump yields a vacuum that is too high and irregular. The surface of the gel is accessible to reagents during the entire procedure. A diagram of a vacuum blotting chamber is represented in Fig. 40.

Olszewska E, Jones K. Trends Gen. 4 (1988) 92-94.

Also pressure can be employed; this is also called pressure blotting.

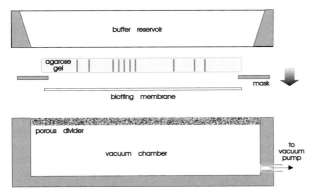

Fig. 40: Transfer of nucleic acids with vacuum blotting in 30 to 40 min.

Vacuum blotting possesses some advantages over capillary blotting, it:

● is faster: 30 to 40 min, instead of overnight;

● is quantitative, there are no back transfers;

● leads to sharper zones and better resolution;

● allows faster depurination, denaturation and neutralization;

in the tank

● reduces the mechanical stress on the gel ;

Gel remains in the tank .

● saves expenses for solutions and paper.

Electrophoretic blotting

Electrophoretic transfers are mainly used for proteins SDS electrophoresis (Towbin *et al.* 1979; Burnette, 1981). Only in some cases also nucleic acids are transferred with the help of an electric field. Either different samples are applied on a gel and analysed together on the membrane, or the antigen solution is separated across the entire gel width and the membrane is cut into narrow strips for probing in different antibody solutions (e.g. patients sera).

Towbin H, Staehelin T, Gordon J. Proc Natl Acad Sci USA. 76 (1979) 4350-4354.
Burnette WN. Anal Biochem. 112 (1981) 195-203.

Tank blotting

Originally, vertical buffer tanks with coiled platinum wire electrodes fixed on two sides were used. For this technique, the gel and blotting membrane are clamped in grids between filter papers and sponge pads and suspended in the tank filled with buffer. The transfers usually occur overnight.

The buffer should be cooled so that the blotting sandwich does not warm up too much.

Semi-dry blotting

Semi dry blotting between two horizontal graphite plates has gained more and more acceptance in the last few years. Only a limited volume of buffer, in which a couple of sheets of filter paper are soaked, is necessary. This technique is simpler, cheaper and faster and a discontinuous buffer system can be used (Kyhse-Andersen, 1984; Tovey and Baldo, 1987).

Kyhse-Andersen J. J Biochem Biophys Methods. 10 (1984) 203-209.
Tovey ER, Baldo BA. Electrophoresis. 8 (1987) 384-387.

The *isotachophoresis* effect occurs here: the anions migrate at the same speed, so that a regular transfer takes place. A system of graphite plates does not need to be cooled. A current not higher than 0.8 to 1 mA per cm^2 of blotting surface is recommended. The gel can overheat if higher currents are used and proteins can precipitate.

Graphite is the best material for electrodes in semi dry blotting because it conducts well, does not overheat and does not catalyze oxidation products.

The transfer time is approximately one hour and depends on the thickness and the concentration of the gel. When longer transfer times are required as for thick (1 mm) or highly concentrated gels, a weight is placed on the upper plate so that the electrolyte gas is expelled out of the sides. Fig. 41 shows a diagram of a semi dry-blotting setup.

In semi dry blotting as well, several blots can be made simultaneously. They are piled in layers called "trans units".

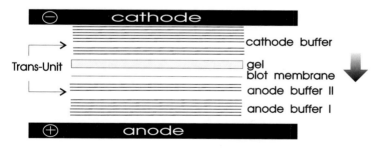

Fig. 41: Diagram of a horizontal graphite blotter for semi dry blotting. Up to six trans units can be blotted at one time (Kyhse-Andersen, 1984).

It is also possible to perform electrophoretic transfers on two membranes simultaneously: *Double Replica Blotting* (Johansson, 1987). An alternating electric field is applied on a blotting sandwich with a membrane on each side of the gel with increasing pulse time, so that two symmetrical blots result.

Johansson K-E. Electrophoresis. 8 (1987) 379-383.

Blotting of gels supported by films

Ready-made or self-made gels backed by support films are used more and more often for electrophoresis of proteins and electrofocusing. These films which are inpermeable to current and buffer must be separated from the gels so that electrophoretic transfers or capillary blotting can be carried out. To separate the gel and the film without damage an apparatus exists with a taut thin steel wire which is pulled between them (see Fig. 42).

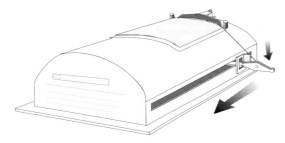

Fig. 42: Instrument for complete and trouble-free separation of gels backed by films.

4.3 Blotting membranes

● Nitrocellulose is the most commonly used membrane. It is available in pore sizes from 0.05 µm to 0.45 µm. The pore size is a measure of the specific surface: the smaller the pores, the higher the binding capacity.

Disadvantages: limited binding capacity and poor mechanical stability.

Proteins adsorbed on nitrocellulose can be reversibly stained so that the total protein can be estimated before specific detection (Salinovich and Montelaro, 1986).

Salinovich O, Montelaro RC. Anal Biochem. 156 (1986) 341-347.

It is also possible to render the blots totally transparent by an imbedding technique.

see the instructions in the second part.

Nitrocellulose is occasionally also used for preparative methods and the proteins can be eluted out again (Montelaro, 1987).

Montelaro RC. Electrophoresis. 8 (1987) 432-438.

A better adsorption of glycoproteins, lipids and carbohydrates is obtained by ligand precoating of nitrocellulose (Handmann and Jarvis, 1985).

Handmann E, Jarvis HM. J Immunol Methods. 83 (1985) 113-123.

● Polyvinylidenedifluoride (PVDF) membranes on a Teflon base possess a high binding capacity and a high mechanical stability like nylon membranes. PVDF membranes can also be used for direct protein sequencing (Matsudaira, 1987).

Disadvantages: staining is not reversible. Matsudaira P. J Biol Chem. 262 (1987) 10035-10038.

● *Diazobenzyloxymethyl* and *diazophenylthioether* papers (DBM, DPT), which must be chemically activated before use, enable a two step binding with molecules: electrostatic and covalent.

DBM and DPT are more and more often replaced by nylon membranes.

● *Nylon membranes* possess a high mechanical stability and a high binding capacity, usually due to electrostatic interactions. This means that staining can be a problem because small molecules are also strongly bound.

Positively and negatively charged but also neutral nylon membranes exist.

Fixing with glutaraldehyde after transfer is recommended to increase the binding of low molecular weight peptides to nylon membranes (Karey and Sirbasku, 1989).

Karey KP, Sirbasku DA. Anal Biochem. 178 (1989) 255-259.

● *Ion-exchange membranes*, diethylaminoethyl (DEAE) or carboxymethyl (CM) are used for preparative purposes because of the reversibility of the ionic bonds.

Disadvantage: These membranes are often very brittle.

● *Activated glass fiber membranes* are used when blotted proteins are directly sequenced. Several methods to activate the surface exist: for example, bromocyanide treatment, derivatization with positively charged silanes (Aebersold *et al.* 1986) or hydrophobation by siliconation (Eckerskorn *et al.* 1988).

Aebersold RH, Teplow D, Hood LE, Kent SBH. J Biol Chem. 261 (1986) 4229-4238.

Eckerskorn C, Mewes W, Goretzki H, Lottspeich F. Eur J Biochem. 176 (1988) 509-519.

Unfortunately a blotting membrane which binds 100% of the molecules does not exist yet. For example, during electroblotting small proteins often migrate through the film while larger proteins have not yet completely left the gel. It is therefore necessary to try and obtain as regular transfers as possible.

4.4 Buffers for electrophoretic transfers

A. Proteins
Tank blotting

A Tris-glycine buffer of pH 8.3 is usually used for tank blotting of SDS gels (Towbin *et al.* 1979). This moderate pH should cause little damage to the proteins. 20% methanol is often added to the buffer to increase the binding capacity of the blotting membrane and to prevent the gel from swelling. At this relatively low pH value the proteins are not highly charged so the transfers are time-consuming and the buffer heats up gradually.

Sometimes to speed up the transfer, ready-stained proteins for blotting and 0.02 to 0.1% SDS to keep hydrophobic proteins in solution are added to the buffer. But this can lead to problems with the binding capacity of the membrane.

Since SDS, methanol, heat and long transfer times are also harmful to proteins, it can be an advantage to use buffers with higher pH values, this saves a considerable amount of time. For example titrate 50 mmol/L CAPS to pH 9.9 with sodium hydroxide.

This transfer takes about 50 min so the buffer does not have time to warm up.

Acid gels with basic proteins and isoelectric focusing are blotted in a tank with 0.7% acetic acid (Towbin *et al.* 1979).

The proteins then migrate towards the cathode.

For glycoproteins, polysaccharides and also lipopolysaccharides 10 mmol/L sodium borate pH 9.2 (Reiser and Stark, 1983) is recommended since boric acid binds to the sugar moieties and thus, in a basic medium, the molecules acquire a negative charge.

Reiser J, Stark GR. Methods Enzymol. 96 (1983) 205-215.

Semi dry blotting

Up to now a continuous Tris-glycine-SDS buffer was often recommended for horizontal blotting (semidry or graphite plate blotting). Yet experience has shown that in general a *discontinuous buffer system* is preferable since it yields sharper bands and more regular and efficient transfers.

A continuous buffer must of course be used when it is necessary to carry out a double replica electroblotting with the semi dry technique.

Continuous buffer:

48 mmol/L Tris, 39 mmol/L glycine, 0.0375% (w/v) SDS, 20% methanol.

Discontinuous buffer system (acc. to Kyhse-Andersen,1984):

Anode I:	0.3 mol/L Tris, 20% methanol
Anode II:	25 mmol/L Tris, 20% methanol
Cathode:	40 mmol/L 6-aminohexanoic acid
	(*same as* ε-aminocaproic acid),
	20% methanol, 0.01% SDS

This buffer system can be used for SDS as well as for native and IEF gels.

If the transfer efficiency of high molecular weight proteins (80 kDa) is not satisfactory, the gel can be equilibrated in the cathode buffer for 5 to 10 min before blotting.

This is mainly applicable to SDS gels.

For enzyme detection the buffer must not contain any methanol, otherwise biological activity is lost. Brief contact with a small amount of SDS does not denature the proteins.

For transfers from urea IEF gels the urea should first be allowed to elute out of the gel by soaking it in cathode buffer. Otherwise the proteins, which possess no charge after IEF cannot bind to SDS as required for electrophoretic transfer.

Caution: proteins can diffuse rapidly out of IEF gels.

B. Nucleic acids

Tank blotting

Acid buffers are often used for DNA blotting: 19 mmol sodium phosphate, 54 mmol/L sodium citrate pH 3.0 (Smith *et al.* 1984).

Smith MR, Devine CS, Cohn SM, Lieberman MW. Anal Biochem. 137 (1984) 120-124.

Semi-dry blotting

For *neutral blotting* 10 mmol/L Tris-HCl, 5 mmol/L sodium acetate, 0.5 mmol/L EDTA, pH 8.7 are used. For *alkaline blotting* 0.4 mol/L NaOH is used (Fujimura et al. 1988). Both these techniques can be carried out in tanks with equal efficiency. However, alkaline blotting damages the plastic material of the tank while graphite plates are resistant to sodium hydroxide.

Fujimura RK, Valdivia RP, Allison MA. DNA Prot Eng Technol. 1 (1988) 45-60.

4.5 General staining

It is often desirable to check the electrophoresis and/or the results of the transfer overall before carrying out a precise evaluation.

Nucleic acids are in general visualized with ethidium bromide which is often added to the gel before separation. The nucleic acids can then be seen under UV light.

For *proteins,* besides staining with Amido Black or Coomassie Brilliant Blue, mild staining methods such as the very sensitive Indian Ink method (Hancock and Tsang, 1983) exist as well as reversible ones with Ponceau S (Salinovich and Montelaro, 1986) or Fast Green FCF (see part II, method 9). The sensitivity of Indian Ink staining and the antibody reactivity of the proteins can be enhanced by alkaline treatment of the blotting membrane (Sutherland and Skerritt, 1986).

Hancock K, Tsang VCW. Anal Biochem. 133 (1983) 157-162.
Sutherland MW, Skerritt JH. Electrophoresis. 7 (1986) 401-406.
Kittler JM, Meisler NT, Viceps-Madore D. Anal Biochem. 137 (1984) 210-216.
Moeremans M, Daneels G, De Mey J. Anal Biochem. 145 (1985) 315-321.

In many cases one can also use:

● a general immunostain (Kittler *et al.* 1984),

● colloidal gold (Moeremans *et al.* 1985),

● autoradiography,

● fluorography (Burnette, 1981).

Nylon membranes bind anionic dyes very strongly so normal staining is not possible but nylon membranes can be stained with caccodylate iron colloid (FerriDye) (Moeremans *et al.* 1986).

Moeremans M, De Raeymaeker M, Daneels G, De Mey. J. Anal Biochem. 153 (1986) 18-22.

4.6 Blocking

Macromolecular substances which do not take part in the visualization reaction are used to block the free binding sites on the membrane.

Denhardts buffer is used for *nucleic acids* (Denhardt, 1966): 0.02% BSA, 0.02% Ficoll, 0.02% polyvinylpyrollidene, 1 mmol/L EDTA, 50 mmol/L NaCl, 10 mmol/L NaCl, 10 mmol/L Tris-HCl pH 7.0, 10 to 50 mg heterologous DNA per mL.

Denhardt D. Biochem Biophys Res Commun. 20 (1966) 641-646.

A number of possibilities exist for *proteins*, 2% to 10% bovine albumin is used most often (Burnette, 1981). The cheapest blocking substances and the ones that cross-react the least are: skim milk or 5% skim milk powder (Johnson *et al.* 1984), 3% fish gelatin, 0.05% Tween 20. The blocking step is quickest and most effective at 37 °C.

Johnson DA, Gautsch JW. Sportsman JR. Gene Anal Technol. 1 (1984) 3-8.

4.7 Specific detection

Hybridization

● *Radioactive probes*

A higher detection sensitivity can be obtained for the analysis of DNA fragments with radioactive DNA or RNA probes which bind to complementary DNA or RNA on the blotting membrane (Southern, 1975; Alwine *et al.* 1977).

Radioactive marking is mainly used for the evaluation of RFLP analysis.

● *Non-radioactive probes*

There is now a trend to avoid radioactivity in the laboratory, so accordingly the samples can be marked with biotin-streptavidin or dioxigenin.

as for immunoblotting

This method is also used for DNA fingerprinting.

Characterization of individuals in forensic medicine.

Enzyme blotting

The transfer of native separated enzymes on to blotting membranes has the advantage that the proteins are fixed without denaturation and thus do not diffuse during slow enzyme-substrate reactions and the coupled staining reactions (Olsson *et al.* 1987).

Immunoblotting

Specific binding of immunoglobulins (IgG) or monoclonal antibodies are used to probe for individual protein zones after blocking. An additional marked protein is then used to visualize the zones. Once again several possibilities exist:

● *Radioiodinated protein A*

The use of radioactive protein A which attaches itself to specific binding antibodies enables high detection sensitivities (Renart *et al.* 1979). But, ^{125}I-protein A only binds to particular IgG subclasses; in addition radioactive isotopes are now avoided as much as possible in the laboratory.

Renart J, Reiser J, Stark GR. Proc Natl Acad Sci USA. 76 (1979) 3116-3120.

● *Enzyme coupled secondary antibodies*

An antibody to the specific binding antibody is used and it is conjugated to an enzyme. Peroxidase (Taketa, 1987) or alkaline phosphatase (Blake *et al.* 1984) are usually employed as the conjugated reagent.

Taketa K. Electrophoresis. 8 (1987) 409-414.
Blake MS, Johnston KH, Russell-Jones GJ. Anal Biochem. 136 (1984) 175-179.

The ensuing enzyme-substrate reactions have a high sensitivity. The tetrazolium method has the highest sensitivity in the peroxidase method (Taketa, 1987).

● *Gold coupled secondary antibody*

Detection by coupling the antibody to *colloidal gold* is very sensitive (Brada and Roth, 1984): in addition the sensitivity can be increased by subsequent *silver enhancement*: the lower limit of detection lies around 100 pg (Moeremans *et al.* 1984).

Brada D, Roth J. Anal Biochm. 142 (1984) 79-83.
Moeremans M, Daneels G, Van Dijck A, Langanger G, De Mey J. J Immunol Methods. 74 (1984) 353-360.

● *Avidin biotin system*

Another possibility is the use of an amplifying enzyme detection system. The detection results from enzymes which are part of a non-covalent network of polyvalent agents (antibodies, avidin): for example biotin-avidin-peroxidase complexes (Hsu et al. 1981) or complexes with alkaline phosphatase.

Hsu D-M, Raine L, Fanger H. J Histochem Cytochem. 29 (1981) 577-580.

● *Chemiluminescence*

The highest sensitivity without using radioactivity can be achieved with enhanced chemiluminscent detection methods (Laing, 1986).

Laing P. J Immunol Methods. 92 (1986) 161-165.

Immunological detection on blots can be automated, for example with the staining unit of the PhastSystem® (Prieur and Russo-Marie, 1988) .

Prieur B, Russo-Marie F. Anal Biochem. 172 (1988) 338-343.

Lectin blotting

The detection of glycoproteins and specific carbohydrate moieties can be performed with lectins. Visualization is carried out by aldehyde detection or, analogous to immunoblotting, with the avidin-biotin method (Bayer *et al.* 1987).

Bayer EA, Ben-Hur H, Wilchek M. Anal Biochem. 161 (1987) 123-131.

4.8 Protein sequencing

The use of blotting for direct protein sequencing has been a big step forward for protein chemistry and molecular biology (Vandekerckhove *et al.* 1985). Blotting ist mostly performed out of one-dimensional SDS gels or 2D gels (Matsudaira *et al.* 1987; Aebersold *et al.* 1986; Eckerskorn *et al.* 1988; Eckerskorn and Lottspeich, 1989). If the proteins to be sequenced have to be separated by isoelectric focusing, an immobilized pH gradient should be used because carrier ampholytes would interfere with the sequencing signals (Aebersold et al. 1988).
In the latest developments, matrix-assisted laser desorption ionization mass spectrometry (MALDI-MS) has been employed to measure the molecular mass of proteins with high precision (Eckerkorn et al, 1992; Strupat et al. 1994).

Vandekerckhove J, Bauw G, Puype M, Van Damme J, Van Montegu M. Eur J Biochem. 152 (1985) 9-19.
Eckerskorn C, Lottspeich F. Chromatographia. 28 (1989), 92-94.
Aebersold RH, Pipes G, Hood LH, Kent SBH. Electrophoresis. 9 (1988) 520-530.
Eckerskorn C, Strupat K, Karas M, Hillenkamp F, Lottspeich F. Electrophoresis 13 (1992) 664-665.
Strupat K, Karas M, Hillenkamp F, Eckerskorn C, Lottspeich F. Anal Chem. 66 (1994) 464-470.

The reference by Simpson *et al.*(1989) presents a review of the different methods.

Simpson RJ, Moritz RL, Begg GS, Rubira MR, Nice EC. Anal Biochem. 177 (1989) 221-236.

4.9 Transfer problems

● *Poor solubility*, especially of hydrophobic proteins, can prevent a transfer. In such cases the blotting buffer should contain a detergent, for example SDS. The addition of urea can also increase the solubility.

The addition of 6 to 8 mol/L urea is only reasonable in semi dry blotting (small buffer volume).

● During *native immunoblotting* according to Bjerrum *et al.* (1987), the electrophoresis gel contains non-ionic detergents. Its binding capacity is hindered in case of direct contact with the blotting membrane. This can be prevented by inserting a 2 to 3 mm thick agarose gel layer containing transfer buffer but no detergent between the gel and the blotting membrane. Nitrocellulose with increased binding capacity can also be used.

Bjerrum OJ, Selmer JC, Lihme A. Electrophoresis. 8 (1987) 388-397.

● High molecular weights cause a slower migration out of the gel. But when blotting is carried out for a long time and/or at high field strengths, low molecular weight proteins detach themselves again from the membrane and are lost. Several possibilities to obtain regular transfers over a wide molecular weight spectrum exist:

● The use of pore gradients in SDS-PAGE,

After the separation the proteins are distributed according to their molecular weight in areas with small or large pores.

● the use of a discontinuous buffer system for semi dry blotting,

The isotachophoresis effect induces a regular velocity.

● the treatment of high molecular weight proteins with protease after electrophoresis,

Limited proteolysis usually does not damage the antigenicity.

● the use of a buffer with another pH,

This increases the mobility.

● the use of a buffer without methanol,

The pores become larger when the gel swells.

● the addition of SDS (0.01 to 0.1%) to the buffer,

Caution: too much SDS reduces the binding capacity.

● the use of native immunoblotting (Bjerrum etal. 1987),

Gels with large pores can be used here,

● blotting for a longer time and placing a second blotting membrane behind.

to trap low molecular weight proteins.

Applications of protein blotting

A series of review articles has been published:

Gershoni JM, Palade GE. Anal Biochem. 112 (1983) 1-15.
Beisiegel U. Electrophoresis. 7 (1986) 1-18.
Bjerrum OJ. Ed. Paper symposium protein blotting. Electrophoresis. 8 (1987) 377-464.
Baldo BA, Tovey ER. Ed. Protein blotting. Methodology, research and diagnostic applications. Karger, Basel (1989).
Baldo BA. In Chrambach A, Dunn M, Radola BJ. Eds. Advances in Electrophoresis 7. VCH, Weinheim (1994) 409-478.

5 Instrumentation

The equipment for capillary electrophoresis and automated DNA sequencing has already been described together with the method.

In most laboratories using electrophoresis the apparatus consists of three principle pieces of equipment:

This means gel electrophoresis in the wide sense.

a) power supply

b) cooling or heating thermostat

c) separation chamber with combined gel casting system.

Concerning a):
For electrophoresis DC power supplies are needed, which yield high voltages and allow to set the maximum output of current, voltage and power.

Concerning b):
Many home-made systems are used without cooling or heating. Yet it has been proved that better and more reproducible separations are obtained with temperature controlled equipment.

Concerning c):
The core of electrophoretic equipment is the separation chamber. A number of types exist because of the many different methods and modifications.

5.1 Current and voltage conditions

To establish electrophoretic separation conditions a few physical rules should be recalled:

This is also important when working under defined conditions.

The driving force behind electrophoresis is the product of the charge $\Theta\pm$ (net charge) of a substance and the electric field E, measured in V/cm. For the speed of migration of a substance v in cm/s this means:

The net charge $\Theta\pm$ can be taken to be the sum of the elementary charges, measured in A s.

$$v = \frac{\Theta\pm \times E}{R}$$

Thus a certain field strength is necessary for an electrophoretic migration.

The frictional constant R is dependent on the molecular radius, r (Stokes radius) in cm and the viscosity, η, of the separation medium measured in N s/cm^2.

To reach the field strength the voltage U must must be applied, it is measured in volt (V) and the separation distance d in cm.

Voltage = field strength × separation distance. $U = E \times d$

If an electric field is applied to a conducting medium (buffer), an electric current, I, will flow. It is measured in amperes (A) yet is usually given in mA for electrophoresis. The magnitude of the current depends on the ionic strength of the buffer. In electrophoresis relatively high currents are used while for isoelectric focusing they are smaller because the pH gradient has a relatively low conductivity.

The product of the voltage and the current is the power, P, given in watt (W):

Power = voltage × current $P = U \times I$

The product of (electrical) power and time is energy. During electrophoresis most of the electrical energy is transformed into heat.

Joule heat

For this reason the temperature should be controlled during electrophoresis. Since the cooling efficiency, that is the heat dissipation, cannot be increased indefinitely, a certain intensity should not be exceeded.

If this is not taken into consideration, the gel can burn through.

Fig. 43 demonstrates the relationship between voltage, current, power and the dimension of the electrophoretic medium. The longer the separation distance, the higher the current necessary to reach a specific field strength. At a given ionic strength, the field strength is proportional to the cross-section: the thicker the gel, the greater the current. The power is proportional to the volume of the gel.

Guidelines for cooled 0.5 mm thick horizontal gels:
Electrophoresis: ca. 2 W/mL gel volume;
IEF: ca.1 W/mL gel volume
IPG: see instructions

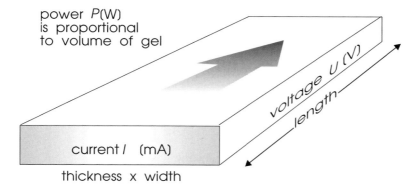

Fig. 43: Schematic diagram of the relationships between the separation medium and current, voltage and power conditions during electrophoresis.

This also means that the power and the current must be reduced if only part of a gel is used, but, for the same separation distance:

    Half gel: *half* the current – *half* the power – *same* voltage.

This is often used for horizontal techniques.

One should always be aware, that the values set in the power supply are maximum values; the real values change during the run, because they are controlled by the conductivity of the buffer and the gel. It is often forgotten, that more concentrated gels have higher resistances than gels with low *T* values.

The conductivity of the system changes during the run, particularly in disc electrophoresis and isoelectric focusing experiments.

5.2 Power supply

Different models and degrees of specification exist:

1. Simple power supplies can be regulated by the voltage.

usually 200 V maximum

2. Typical electrophoresis power supplies can be run with constant current or constant voltage.

usually up to 500 or 1000 V, 200 or 400 mA

3. Power supplies which are also designed for isoelectric focusing supply high voltages. Their power is also stabilized so that a maximum setting can be programmed.

usually t up to 3000 or 5000 V, 150 or 250 mA, 100 or 200 W

In isoelectric focusing the conductivity of the gel drops when the pH gradient has formed, and the buffer ions have migrated from the gel. Regulating the maximum power prevents overloading the gel with high voltages. Additional control over the focusing conditions is provided by a volt/hour integrator.

4. Programmable power supplies have an additional microprocessor with which different separation conditions with various steps can be recalled.

A volt/hour and an ampere/hour integrator are usually also built in.

For example:
First phase: low voltage for gentle sample entry into the separation medium.
Second phase: high voltage for rapid separation and to avoid diffusion.
Third phase: low voltage to avoid diffusion and migration of the zones.

For long separation distance gels, high voltages must be applied and low current. Optimal resolution and high reproducibility is obtained, when these gels are run with a "Voltage ramping".

In method 11 on page 241 a voltage ramping program is described .

5.3 Separation chambers

5.3.1 Vertical apparatus

Electrophoresis is carried out in vertical cylindrical gel rods in glass tubes or gel slabs which are cast in glass holders. The samples are applied on the surface of the rods or in gel pockets with a syringe. The current is conducted through platinum electrodes which dip in the buffer tanks.

An example of a vertical system with temperature regulation which can be used for gel slabs or cylindrical gel rods is shown in Fig. 44. To dissipate Joule heat the bottom buffer is cooled by a heat exchange system. For classical 2D electrophoresis (see Fig. 24), both separations are performed in an apparatus like this. When high protein amounts have to be loaded for protein sequencing, horizontal isoelectric focusing in immobilized pH gradients is used in the first dimension with vertical SDS electrophoresis in 1.5 mm thick gels in the second dimension (Ekkerskorn *et al.* 1988; Hanash *et al.* 1991; Bjellqvist *et al.* 1993).

Hanash SM, Strahler JR, Neel JV, Hailat N, Melham R, Keim D, Zhu XX, Wagner D, Gage DA, Watson JT. ProcNatlAcadSciUSA. 88 (1991) 5709-5713

Bjellqvist B, Sanchez J-C, Pasquali C, Ravier F, Paquet N, Frutiger S, Hughes GJ, Hochstrasser D. Electrophoresis. 14 (1993) 1375-1378

In the *IsoDalt* equipment designed by Anderson and Anderson (1978) ten separations can be run parallely.

Directions for vertical electrophoresis can be found in the book by Andrews (1986), the Hoefer Applications guide (1994) and in method 15 of this book.

Hoefer Protein Electrophoresis Applications Guide (1994) 18-54.

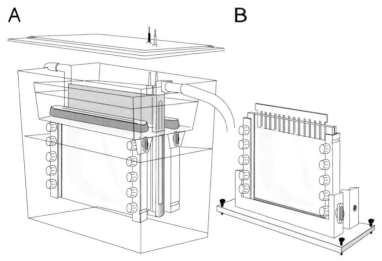

Fig. 44: Vertical electrophoresis chamber.
(A) Separation chamber with heat exchanger, up to 4 gels can be run in parallel;
(B) Gel casting equipment with comb for sample wells.

Sequencing chambers

Since long gels are required for manual DNA sequencing, special chambers with thermostatable heating plates, are used (Fig. 45). The ultrathin large gels are prepared with the horizontal sliding technique of Ansorge and De Maeyer (1980). The same setup is employed for automated sequencing, however, because of the on-line detection of the fragment, the gels can be shorter.

Ansorge W, De Maeyer L. J Chromatogr. 202 (1980) 45-53.

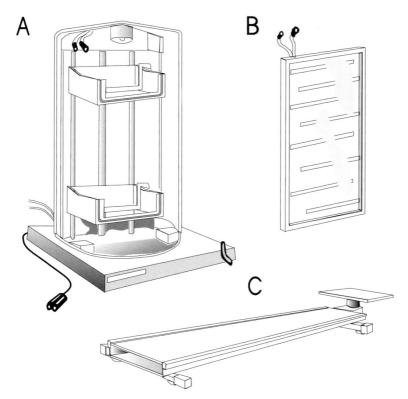

Fig. 45: DNA sequencing chamber.
(A) Vertical chamber with safety cabinet;
(B) Thermostatable plate;
(C) Horizontal gel casting apparatus for the sliding technique.

5.3.2 Horizontal apparatus

DNA analysis in agarose gels

For analytical and preparative separation of DNA fragments and RNA restriction fragments *"submarine" chambers* are usually used. The agarose separation gel is submerged under a thin layer of buffer between the lateral buffer tanks.

Horizontal chambers exist in different sizes.

For electrophoresis in a *pulsed field* a controlling device is connected to the power supply, which switch the electrodes – at predefined frequency – in the north/south and the east/west directions. Diodes are built-in to the electrodes so that when they are switched off, they cannot influence the field. Since these separations can last for a long time – up to several days – the buffer must be cooled and circulated (Fig. 46).

For non-homogeneous fields the point electrodes are placed in electrode grooves set at right angles.

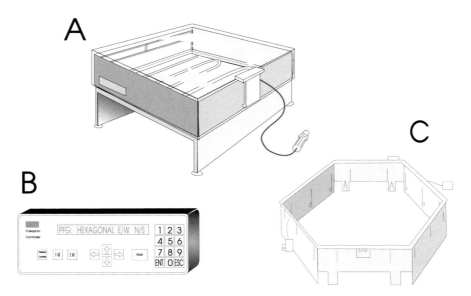

Fig. 46: System for pulsed field DNA gel electrophoresis (PFG). - (A) PFG submarine chamber with cooling coil and buffer circulation pump (not visible); - (B) Programmable pulse controlling device; - (C) Hexagonal electrode for linear sample lanes.

Protein and DNA analysis in polyacrylamide gels

Horizontal chambers with thermostatable plates and lateral buffer tanks are very versatile (Fig. 47): They are equipped for analytical and preparative isoelectric focusing, for several variations of immuno and affinity electrophoresis, all zone electrophoresis techniques in restrictive and non-restrictive gels, high resolution 2D electrophoresis, as well as semi-dry blotting. High voltages can be applied, because there are no problems with insulation of buffer tanks; and many techniques can be applied without using a buffer tank at all.

If the gel supporting area does not have to be too large, thermoelectric cooling or heating with Peltier elements can be used instead of employing a thermostatic circulator (Fig. 48).

5.4 Staining apparatus

Very helpful, particularly for silver staining with its many steps, is a staining apparatus. It saves time and improves reproducibility of the result considerably. Figure 49 shows an automated gel stainer which consists of a programmable controler, a rocking tray, a 10-port valve and a fast peristaltic pump.

and protein analysis in agarose gels

Most of the instructions in part II are designed for this kind of electrophoresis equipment, because almost all methods can be performed on it.

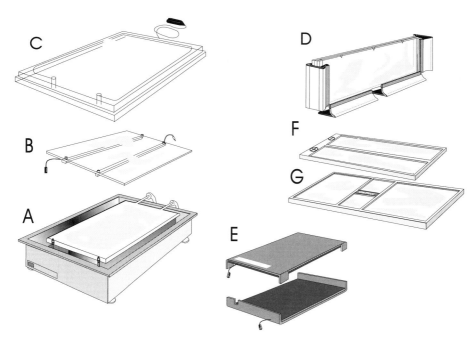

Fig. 47: Horizontal electrophoresis system. - (A) Separation chamber with cooling plate and lateral buffer tanks; - (B) Electrodes for isoelectric focusing and electrophoresis with buffer strips; - (C) Safety lid; - (D) Gel casting cassette for ultrathin homogeneous and gradient gels; - (E) Graphite plate electrodes for semi-dry blotting; - (F) PaperPool for soaking of electrode wicks;- (G) GelPool for rehydration of dry gels.

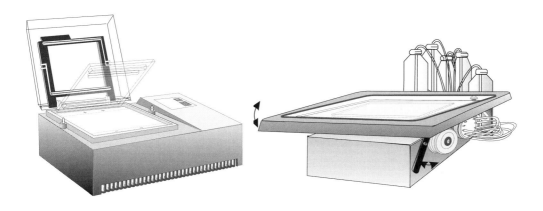

Fig. 48: Horizontal apparatus with Peltier cooling and built-in electrode system.

Fig. 49: Automated gel stainer for gel types of different sizes with and without film support.

5.5 Automated electrophoresis

A complete automated electrophoresis system, the PhastSystem®, is composed of a horizontal electrophoresis chamber with a Peltier element which cools and heats the separation bed, and a built-in programmable power supply and development unit (see Fig. 50). The current and temperatures for separation and staining as well as the different development procedures can be programmed and recalled for the various electrophoresis and staining methods. There is a blotting unit with graphite electrodes for electrophoretic transfers.

The gels only needs to be placed in the separation compartment, the sample applied and the gel transferred to the development chamber after the separation.

Special gels with support films for focusing, titration curves and electrophoresis, native and SDS-electrophoresis buffers, which are cast in agarose, as well as staining tablets and a silver staining kit exist for this system. The gels and buffer strips can also be handmade (Pharmacia,1988)

Pharmacia LKB Instructions for the preparation of gels for the PhastSystem® (1988).

The samples are applied automatically - at a defined time - with multiple sample application combs. The separation and development steps occur very quickly because the gels are only 0.3 and 0.4 mm thick and relatively small, 4 cm × 5 cm.

These combs correspond to sample applicators.

A SDS-electrophoresis run in a pore gradient gel lasts 1.5 h including silver staining. Therefore,very sharp bands are obtain, and in spite of the small geometry, a high resolution is achieved.

The development unit can be warmed to 50 °C. It contains tubing for the entry and exit of the staining solutions and a membrane pump for creating vacuum or pressure in the chamber as well as for emptying it or pumping in solutions. There also is a gel holder for rotating the gel in the solution. All the functions can be programmed. The timer, temperature and level sensors regulate the execution of the program.

The development unit is monitored by the separation and control unit.

The advantages of such an automated electrophoresis and development unit are numerous:

● rapid and very reproducible separations,

● electrophoresis can be carried out immediately and without preparation,

● problems due to "human error" are reduced to a minimum,

● reduction of the work load (ready-to-use gels and buffers, automated separation and development),

● multiple sample application,

● no handling of liquid buffers,

- clean work, can be carried out at any location,
- easy and rapid switching between separation and development methods,
- practical presentation possibilities by mounting the stained gels in slide frames.

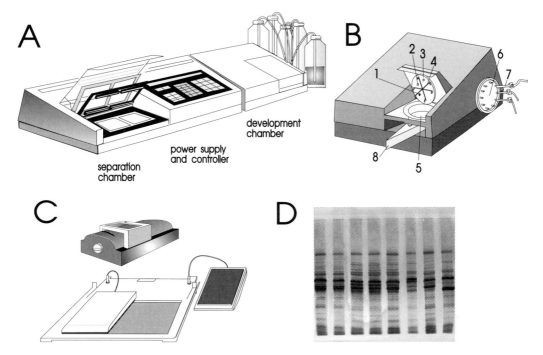

Fig. 50: System for automated electrophoresis (PhastSystem®). - (A) The complete apparatus; (B) the development unit: (1) opening to the membrane pump, (2) temperature sensor, (3) level sensor, (4) rotating gel holder, (5) development chamber, (6) 10-port valve, (7) numbered PVC tubing, (8) closing mechanism; (C) Blotting system to place in the separation compartment and filmremover. (D) PhastGel medium: SDS electrophoresis after automated silver staining (By kind permission of Professor A. Görg).

5.5 Safety measures

In most electrophoresis techniques high voltages (>200 V) are used to reach the field strengths required for separation. So as not to threaten safety in the laboratory, electrophoresis should only be performed in closed separation compartments. A danger of electric shock by contact exists when the separation system is open. In addition, the system should shut off automatically in the event of a short-circuit.

The cables and plugs must be designed for constant current with high voltages.

Many separation chambers, e.g. for vertical electrophoresis or submarine techniques, are only licensed for voltages up to 500 V. Power supplies with specifications for versatile uses can be equipped with a sensor for coded plugs, which only allows the maximum voltages for special chambers.

*Versatile uses means: **either** high voltages **or** high currents. In any case, the power is limited by the heat production.*

The separation unit should be designed in that way, that the the plugs and sockets placed so that the current is automatically shut off if the chamber is opened by mistake. Electrophoresis systems should be designed so that the power supply is above or at least at the same height as the separation unit, to avoid buffer running into the power supply should it spill out.

Electrophoresis equipment should be kept in a dry place.

5.7 Environmental aspects

Choosing an appropriate instrument system can also be influenced by environmental aspects.

Using buffer strips or buffer wicks is advantageous over large buffer volumes for two main reasons:

only feasible with horizontal systems.

● No excess chemicals: Only that amount of chemical is used which is needed for the run of the number of samples to be separated.

The gels, strips or wicks can be cut to size.

● The volume of materials and liquid disposed is much less than for buffer tanks in the conventional techniques.

This applies for chemicals as well as for radioactivity.

Also the detection method can influence both the choice of the appropriate electrophoresis method and the instrumental system.

Here are two examples:

● *DNA electrophoresis:* As ethidium bromide is not welcome in every laboratory, the agarose-submarine-ethidiumbromide technique is – in many cases – replaced by using rehydrated thin polyacrylamide gels on carrier films with subsequent silver staining.

There are a few more advantages with the alternative polyacrylamide gelmethod; see Method 11 and the following on page 233 of this book.

● *Isoelectric focusing:* In isoelectric focusing experiments, efficient fixing of the proteins and simultaneous washing out of the carrier ampholytes can only be done with trichloroacetic acid. Many laboratories with routine applications and a high gel throughput look for an alternative for this halogenated acid. Some separation problems may be solved by employing a basic or acidic native electrophoresis technique instead, because alternative fixing and staining methods can easily be used here.

The technique of native electrophoresis in washed and rehydrated polyacrylamide gels offers a number of possibilities, see method 5 in this book.

6 Interpretation of electropherograms

6.1 Introduction

6.1.1 Purity control

To control purity, electrophoretic methods are usually used in combination with chromatography. The physico-chemical properties of the substances to be investigated are generally known.

Gel electrophoresis has the advantage that a large number of samples can be analyzed at one time.

SDS electrophoresis is the most frequently used technique for protein analysis. During SDS electrophoresis, configurational differences and irrelevant charge heterogeneities of polypeptides or enzymes are eliminated so that only real impurities appear as extra bands of different molecular weights. For analysis of low molecular weight peptides special designed gels or buffer systems are usually applied. In some cases the load of the major fraction has to be very high, because low amounts of contaminating proteins must be detected and – sometimes – quantified.

Depending on the type and degree of impurity, staining methods with high or low sensitivity are employed; blotting techniques are used for many experiments.

In cases of very high protein loads, a complete discontinuous buffer system has to be applied (see page 29).

When the degree of glycosylation or the electrical properties of a protein are significant, isoelectric focusing is used. In some cases band heterogeneity caused by the various conformations of a molecule must be taken into account when interpreting results.

Agarose submarine gels are often used for nucleic acids. Ethidium bromide staining is used for detection, either with hybridization in the gel or after blotting on an immobilizing membrane.

6.1.2 Quantitative measurements

Quantification is either carried out with direct UV measurement of the zones in capillary systems or - when support materials such as gels or films are used - indirectly by autoradiography or staining of the zones followed by densitometric measurement.

In *capillary electrophoresis*, the scans resemble chromatograms and the peaks can be integrated as in standard chromatography or HPLC.

In *immunoelectrophoresis*, the distance between the precipitation line and the origin is a measure of the concentration of the substance in the sample, regardless of whether the process involves electrophoretic or electroosmotic migration or diffusion.

It is, however, important to know the quantity of antibody introduced as well as the titer of the antigen-antibody mixture. The clearest and most reproducible results are obtained with the rocket method of Laurell (1966). The areas enclosed by the precipitation lines are proportional to the amount of antibody in the sample. In many cases, it is accurate enough to simply measure the height of the precipitin arc.

The quality of the results depends on the quality of the antibodies: impurities with cross-reacting antibodies must be eliminated.

The success of quantitative determinations using electrophoretic separations on gels or other supports depends on several factors:

● During sample preparation, substance losses through adsorption on membranes or column material during desalting or concentration should be avoided as well as the formation of irreversible deposits during extraction or precipitation. Complexing or chelating agents must also be removed from the sample or else complex formation must be inhibited.

For practical reasons, methods for sample preparation are described in connection with each separation method.

● The application method must be chosen in such a way that all substances completely penetrate the separation medium. This is especially critical during isoelectric focusing of heterogeneous protein mixtures since different proteins are unstable at different pHs or have a tendency to aggregate. In such cases it can be assumed that all proteins will not penetrate the medium at the same point. During sample application, the volumetric precision of the syringe or micropipette is also important.

● The quality of the separation is crucial for densitometric measurements. Wavy distorted zones, which result for example from the salt concentration being too high, lead to questionable densitograms. In addition, a zone can only be properly quantified when it is well separated from the neighboring bands.

Qualitative differences during separations can be compensated with the help of marker proteins.

● The prerequisite for a reliable quantification is an effective staining of the bands while avoiding destaining during background washing.

Hot staining and colloidal staining methods are recommended.

Nevertheless, it should always be presumed that differences in staining effectiveness exist. For this reason a protein mixture consisting of known pure substances (e.g. marker proteins) at various concentrations (serial dilutions) should always be applied and run in parallel as is done in qualitative comparison of bands for molecular weight or isoelectric point determination.

Since each protein possesses a specific affinity to the dye used which is different from other proteins, it should always be remembered that only relative quantitative values can be determined.

For example measurements relative to albumin.

Gradient gels almost always display an increasingly or decreasingly shaded background after staining. This should be accounted for during photometric measurement of the electrophoresis bands and the ensuing integration of the surfaces.

6.2 Computer aided analysis

In many cases it is sufficient to compare the bands and spot patterns visually, to photograph the separations or keep the original gels in dry or humid form. However, measurement and further analysis of the electropherograms is necessary for a number of applications:

e.g. the identification of substances or simple comparison of electrophoresis patterns.

● It is difficult, even impossible to determine the intensity of individual fractions visually. For example:

Homozygous and heterozygous genotypes must be differentiated during genetic investigations. Usually only the presence or absence of a band or spot can be detected visually.

Homozygote means: intense band. Heterozygote means: partial intensity.

When protein metabolism kinetics are studied, the increase or decrease of certain fractions must be recognized..

● Single fractions or groups of fractions should be quantified. This is only possible by scanning the separation traces and integrating the surfaces of the peak diagram with a densitometer. It is necessary to measure the zones as exactly as possible since they can present different forms and zone widths depending on the method used.

A whole series of factors must be taken into account during quantification. They will be discussed later.

● The interpretation of 2D electropherograms with several hundred spots is complicated and time-consuming, which is why computers are essential. The patterns are recorded with a densitometer, a desk top scanner or a video camera.

see 2D electrophoresis, page 31.

● Densitograms are the usual form of representation in many areas such as clinical chemistry for example.

● Electrophoresis patterns can often be compared more exactly by densitometry than by the simple visual comparison.

● Data processing is used more and more often because of the amount of data collected in laboratories. To be able to evaluate, save and process results of electrophoresis by computer, the lanes must be digitalized by a densitometer.

In many laboratories, scanning of gels and subsequent printing with a laser or ink jet printer has replaced photographing.

● Molecular weights or isoelectric points of samples can be assigned by computer.

● For many uses, especially in routine analysis, the separation distances are shortened and the results more difficult to interpret visually. Bands which lie close to one another can be resolved by high resolution scanning and enlarged by the computer.

magnifying glass effect

Before the analysis, the electropherograms have to be fed into a computer with a video camera, a desk top scanner or a densitometer.

6.2.1 Instumentation for image acquisition

Video Camera: The resolution of video cameras has been improved in the last years. They can be used for visible and UV light. Video cameras are mostly employed for the evaluation of two-dimensional electrophoresis and Ethidium bromide stained agarose gels.

CCD cameras provide digital signals. They can be coupled directly to a computer.

Desktop Scanners: High performance instruments are available, which can scan in both, reflectance and transmittance (Fig. 51). They scan very fast, provide very high resolution, and they are much less expensive than densitometers.

Some desk top scanners can be calibrated and used for quantitation..

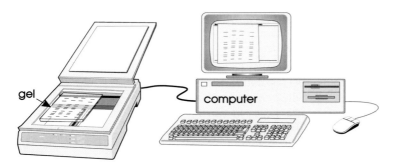

Fig. 51: Desk top scanner for rapid high resolution scanning of one- and two-dimensional electrophoresis gels.

Densitometers: Those are mobile photometers which measure electrophoresis or thin-layer chromatography traces. The R_f value (relative mobility) and the extinction (light absorption) of the individual zones are determined.

The measurement of the separation traces is termed "scanning".

The result is a peak diagram (densitogram). The surfaces under the peaks can be used for quantification. Such densitometers have long been used in routine clinical analysis for the evaluation of cellulose acetate and agarose gels. These densitometers are equipped with specific evaluation programs to aid diagnosis.

The resolution is not very high in these electrophoresis techniques.

The introduction of high resolution gel electrophoresis and blotting methods makes high demands which can only be partly fulfilled by conventional densitometers using white light (Westermeier *et al.* 1988). Substantially higher resolution and sensitivity is achieved, when a laser is employed. A schematic drawing of an apparatus for laser densitometry for high resolution gel electrophoresis is presented in Fig. 52.

Westermeier R, Schickle HP, Thesseling G, Walter WW. GIT Labor-Medizin. 4 (1988) 194-202.

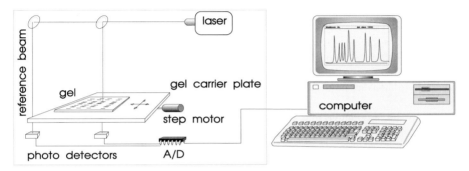

Fig. 52: Apparatus for laser densitometric evaluation of high resolution gel electrophoresis.

Computer: Because today's personal computers have very fast processors, excellent graphic devices, and can be equipped with high capacity hard disks for reasonable prices, it is no longer necessary to connect densitometers and scanners to a work station or a UNIX® based computer. The data can easily be further processed with the desk top publishing softwares.

Nobody must be any more afraid of the computer aided gel evaluation. Personal computers are much easier to operate than work stations. The new scanning and data processing softwares are very convenient and easy to handle.

6.2.2 The optics of a densitometer

When scanning high resolution electrophoresis patterns reliable results are only obtained when the densitometer is able to reproduce the bands or spots on the autoradio/fluorograph or blotting membrane exactly and without distortions. In addition it should yield values which are proportional to the absorption. For this an intense light beam is necessary, a spectral bandwidth as narrow as possible, and a high resolution and a high focusing depth, especially for gels that have not been dried.

Two alternative light sources exist: white light with filters or laser light.

To obtain authentic values by densitometric measurement of stained zones in gels or blackened zones on X-ray films, the measurements must always be carried out in the transmission mode:

● False values can occur as a result of scattering when reflection is measured.

In contrast to reflection measurements, which is the standard procedure in thin layer chromatography, transmission is preferred.

● Reflection only measures the substances which lie at the surface, yet, especially in blotting membranes part of the substances are to be found inside or at the back of the membrane.

When light is shone through a medium, part of it is absorbed. The remaining intensity can be measured by a photoelectric cell.

Each substance has its own absorption spectrum as is known from spectrometry, meaning that the extinction of light changes with the wavelength.

The absorption spectrum is a physical constant.

The remaining light intensity measured depends on the light source, the filter used and the absorption spectrum of the sample. Since the relationship between these parameters is quite complicated, it is easiest to measure at a single wavelength.

White light has the advantage that an optimum wavelength interval can be set with an appropriate filter. A wavelength corresponding to an absorption maximum of the sample can be chosen, and the highest sensitivity lies there. The value measured is not dependent on the wavelength in a narrow range around the peak maximum. If one wants to measure the shoulder of an absorption peak, the bandwidth of the light should be as narrow as possible.

In general the bandwidth of the light used should be narrower than the absorption peak.

Laser light has a fixed wavelength, thus it is not possible to choose an optimum range. However it possesses a very narrow bandwidth, in fact one single spectral line.

Wavelength of the helium-neon-laser: 632.5 nm

Since the wavelength of a laser is a physical constant, like the absorption spectrum of a substance, an optimum reproducibility and good linearity of the values at all points of the absorption spectrum are obtained. Of course, the substance measured must have an absorption different from zero at the wavelength used.

The absorption spectrum of a substance with the typical bandwidths of filtered conventional white light and laser light are represented in Fig. 53.

Optical density: A further prerequisite for correct and linear values is a sufficient light intensity. In high resolution electrophoresis the zones have an optical density (O.D.) up to 4 O.D.

The background of blotting membranes already reaches 2.5 O.D.

The unit O.D. for the optical density is mostly used in biology and biochemistry and is defined as follows: 1 O.D. is the amount of substance, which has an absorption of 1 when dissolved and measured in 1 mL in a cuvette with a thickness of 1 cm.

According to Lambert-Beer's law the extinction of light of a specific wavelength shining through a dissolved substance is proportional to its concentration.

The intensity of the light absorption of a substance is called extinction.

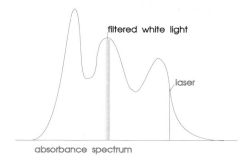

Fig. 53: Schematic diagram of the absorption spectrum of a substance with the bandwidths of filtered white and laser light.

The quantification of results is only possible when the absorptions are linear. With white light this is possible up to 2.5 O.D. at most, and with laser light up to 4 O.D.

The zones must be reproduced exactly to obtain correct results, this means that a high resolution is necessary. The resolution of a densitometer is determined by three parameters: the width of the light beam, the amplitude of the steps of the motor and the field depth. The necessity for a narrow light beam is illustrated in Fig. 54.

Neighboring zones will not be measured down to the baseline if the beam is too wide and the absorption maximum of narrow bands will not be attained.

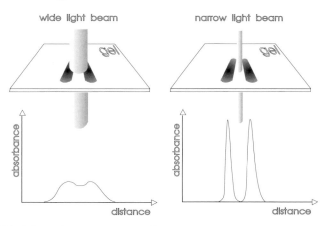

Fig. 54: Schematic representation of the measurement of narrow neighboring bands with a wide and a narrow beam of light. The narrower the beam of light, the more exact the result.

Reducing the width of a beam of white light by making the slit narrower is only possible to a value of about 100 μm, otherwise too much intensity is lost. A laser beam with a width of 50 μm still possesses enough intensity to yield linear values up to 4.0 O.D.

Two-dimensional measurement of bands

For exact measurement it is necessary to take the different lateral expansions of the individual protein bands into account.

Lateral means perpendicularly to the direction of the electrophoretic separation.

There are two possibilities:

- when measuring in one-dimensional mode the width of the electrophoresis trace can be recorded. The beam of light, shifted by a unit each time, then travels over the trace until the whole width is recorded ("X-width");

The individual results are averaged.

- or the whole gel is measured in two-dimensional mode and the corresponding bands are then defined in the computer . Desk top scanners work only in this mode.

In this case the bands are treated like the spots of 2D electrophoresis.

When everything has been determined, the densitogram can be integrated.

6.2.3 Integration and baseline

Gradient gels in particular have backgrounds which are not linear. It is crucial to measure them exactly because the area of a peak in O.D. × mm, is calculated down to the baseline. Fig 55 A shows that errors of up to 50% can occur when a simple horizontal baseline is considered as is often done.

A peak corresponds to a protein band in the absorption diagram.

Appropriate integration software should be able to handle this problem either automatically or manually, and ideally both ways (Fig. 55 B).

A

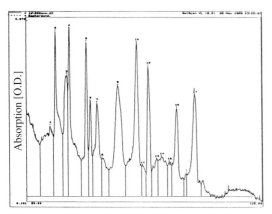

B

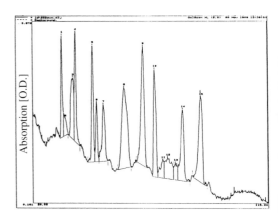

Fig. 55: Densitogram of marker proteins in an SDS pore gradient gel. - (A) The *horizontal* baseline does not take the increasing background staining into consideration; (B) with a *manual* baseline the real peak area can be approached more reliably.

6.2.3 Evaluation of densitograms

Qualitative results

After calibration with marker proteins the R_f values of molecular weights or isoelectric points can be assigned automatically (see Fig.56).

Traditionally, molecular weight determinations are done by recording the values on semi-logarithmic graph paper.

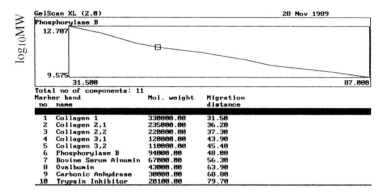

```
GelScan XL (2.0)                          28 Nov 1989
Phosphorylase B
12.707

                        ⊟

 9.575
      31.500                                        87.000
Total no of components: 11
Marker band              Mol. weight   Migration
 no  name                              distance

  1  Collagen 1          330000.00      31.50
  2  Collagen 2,1        235000.00      36.20
  3  Collagen 2,2        220000.00      37.30
  4  Collagen 3,1        120000.00      43.90
  5  Collagen 3,2        110000.00      45.40
  6  Phosphorylase B      94000.00      48.00
  7  Bovine Serum Alnumin 67000.00      56.30
  8  Ovalbumin            43000.00      63.90
  9  Carbonic Anhydrase   30000.00      68.80
 10  Trypsin Inhibitor    20100.00      79.70
```

log₁₀MW label: $\log_{10}MW$

Fig. 56: Molecular weight calibration curve obtained by densitometric measurement of the separation trace of a mixture of markers containing high molecular weight collagen. Staining with Coomassie Blue.

Quantification

Absorption to concentration ratio

It should first be clearly understood that the conventional laws of photometry do not apply to densitometry. While dilute solutions (mmol concentrations) are measured in the UV and visible ranges during photometry, the absorbing medium is a highly concentrated precipitate in gel densitometry. It usually is a protein which is bound to a chromophore. A typical protein curve is shown in Fig. 57.

It is very complicated, if not impossible, to calculate a constant for a protein-dye aggregate.

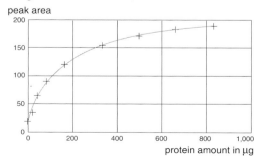

Fig. 57: Color intensity curve of carboanhydrase, separated with SDS-PAGE and stained with colloidal Coomassie Brilliant G-250: Peak area/amount of protein.

Similar problems occur during densitometric evaluation of X-ray films. The Lambert-Beer law cannot be applied to densitometry because it is only valid for very dilute, "ideal" sample solutions.

External standard

An external standard should be used for a series of experiments. If a mixture of proteins is used, the amount of each protein applied will be known.

If the amount of a specific protein is to be calculated in mg, it should be remembered that every protein has a different affinity for a dye , e.g. Coomassie (Fig. 58). The diagram is based on the calculation of the amount of marker protein in the mixture per measured and integrated peak area:

Lactalbumin	138 µg	Carboanhydrase (CA)	108 µg
Trypsin inhibitor	133 µg	Phosphorylase B	60 µg
Ovalbumin	122 µg	BSA	57 µg

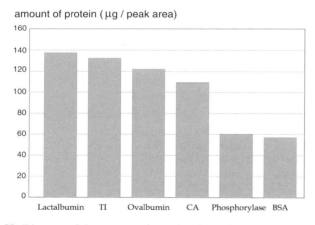

amount of protein (µg / peak area)

Fig. 58: Diagram of the amount of protein of individual markers per integrated peak area. TI: trypsin inhibitor from soybeans, CA: carboanhydrase, BSA: bovine serum albumin. Staining with Coomassie Blue.

Therefore an unknown protein should not be calculated directly using albumin for calibration, the protein should itself be available in the pure form. But if the values are recorded compared to albumin, the correct values can be recalculated later.

Recalculated means: if the protein to be determined should later become available as pure substance, a comparison can then be carried out.

The path to an absolute value lies through the identification of a protein with immuno or lectin blotting and the use of the procedure described above. If there is no antibody or specific ligand available, a subject for a thesis has been found!

Equipment for section II

Almost all methods described here are performed in a horizontal system with the same equipment. In method 16 the procedure for a vertical equipment is explained.

The small items and pieces of equipment as well as the principal stock solutions can be used for almost all the methods.

In principle the sequence of the first ten methods does not correspond to their importance or frequency of use but rather their simplicity and cost. The DNA methods are placed to the end, because all of them are new developments.

This should also help for the planning of an electrophoresis course.

Methods:
Small molecules:
1. PAGE of dyes
4. Native PAGE in amphoteric buffers

Proteins:
2. Agarose and immunoelectrophoresis
3. Titration curve analysis
4. Native PAGE in amphoteric buffers
5. Agarose IEF
6. PAGIEF in rehydrated gels
7. SDS polyacrylamide gel electrophoresis
8. Semi-dry blotting of proteins
9. Isoelectric focusing in IPG
10. High resolution 2D electrophoresis
15. Vertical PAGE

DNA:
11. PAGE of double stranded DNA
12. Native PAGE of single stranded DNA
13. Denaturing gradient gel electrophoresis
14. Denaturing PAGE of DNA
15. Vertical PAGE

Instrumentation

Multiphor II	horizontal electrophoresis unit	*for methods 1 to 14*
Mighty Small SE 250	vertical electrophoresis unit	*only for method 15!*
EPS 3500 XL	programmable power supply 3500 V	
Multitemp II	thermostatic circulator	
Automated Gel Stainer	silver staining apparatus	
Novablot	graphite plate electrodes for blotting	

Film Remover   apparatus for removing support films
IPG strip kit   for 2D electrophoresis
Ultroscan XL   laser densitometer
Desk top scanner
Computer (Windows)
Imagemaster   1 D and 2D evaluation software
Universal gel kit: contains glass plates, glass plates with
                          gaskets, clamps, gradient mixer, compensating stick, tubing,
                          tubing clamps, scalpel, tape
GelPool   tray for rehydration of dry gels
PaperPool   double tray for soaking of electrode wicks

Immuno-
electrophoresis kit: contains levelling table, spirit level, glass
                          plates, electrode wicks, holder, special
                          scalpell, punching template, gel punch with vacuum tubing,
                          sample application film

Field strength probe and voltmeter.
Hand roller, water gauge,
silicone rubber sample applicator mask;
humidity chamber for the agarose techniques,
stainless steel staining trays,
destaining tanks,
glass tray for silver staining

Scissors, spatulas
Assorted glass ware: beakers, measuring cylinders, erlen-
meyers, test tubes etc.
Magnetic stirrer bars in different sizes
Graduated pipettes of 5 and 10 mL + pipetting device
(e.g. Peleus ball)
Micropipetttes adjustable from 2 to 1,000 µL

Consumables:

Disposible gloves, tissue paper, filter paper, Scotch tape,
pipette tips, Eppendorf cups,
test tubes with screw caps 15 and 50 mL

SPECIAL LABORATORY EQUIPMENT

Method	1	2	3	4	5	6	7	8	9	10	11	12	13	14	15
Glass rod		○			○										
Heating block							□					□		□	
Heating cabinet or incubator		○			○		○	○	○				○		○
Heating stirrer			■	■		■	■	■							■
Laboratory elevator ("Laborboy")							○		○				○		○
Microwave oven		○			○										
Paper cutter "Roll and Cut".										○					
Rocking platform							■	■	○	○	○	○	■	■	■
Small magnetic stirrer							○	○	○	○	○	○	○	○	○
Spatula in different sizes	○	○	○	○	○	○	○	○	○	○	○	○	○	○	○
Table centrifuge	□	□	□	□	□	□	□	□	□	□	□	□	□	□	
Forceps straight and curved						○	○	○	○	○	○		○	○	
UV lamp									■						
Ventilator		○	○	○						○	○	○	○	○	○
Watch glasses		○	○		○										
Water jet vacuum pump		○	○			○				○					

□ For sample preparation, ○ for the method, ■ for detection

CONSUMABLES

Method	1	2	3	4	5	6	7	8	9	10	11	12	13	14	15
Blotting membrane, nitrocellulose								○							
Dymo tape			○	○			○				○	○	○	○	
Electrode wicks	○		○	○			○				○	○	○	○	
Filter paper	○	○ ■	○	○	○ ■	○	○	○	○	○	○	○	○	○	
Focusing strips					○	○		○	○	○					
GelBond® Film for agarose (12.5 × 26 cm)		○		○											
GelBond® PAG Film (12.5 × 26 cm)			○				○		○			○	○		
GelBond® PAG Film (20.3 × 26 cm)										○					
Parafilm®	○			○								○			○
Pipette tips	□○	□○	□○	□○		□○	□○		□○		□○	□○	□○	□○	□○
Plastic bags	○		○			○	○		○		○	○	○	○	○
Polyester film untreated	○		○			○	○		○		○	○	○	○	
PVC film (overhead)								■							
Sample application pieces						○	○								

□ For sample application, ○ for the method, ■ for detection

CHEMICALS

Method	1	2	3	4	5	6	7	8	9	10	11	12	13	14	15
Acetic acid (96 %)		■	■	□○	■○	■	■	○	■○	■○	■○	■○	■○	■○	■
Acrylamide	○		○	○	○	○	○		○	○	○	○	○	○	○
Activated carbon, granulated			■	■		■	■								
Agarose L		○													
Agarose IEF					○										
Mxed bed ion-exchanger	○	○	○	○	○	○			○	○	○	○	○	○	
ε-Aminocaproic acid				○				○							
Ammonium nitrate			■	■	■	■			■						
Ammonium persulphate	○		○	■	■	○	○		○	○	○	○	○	○	○
Ammonium sulphate							■								
Ampholine® carrier ampholyte, depending on the pH interval			○		○	○				□					
Antibodies		○			■		○	■	○	○	○	○	○	○	○
Benzoinmethyl ether								■							
Bisacrylamide	○		○	○	○	○	○		○	○	○	○	○	○	○
Boric acid											○	○	○	○	
Calcium lactate		○													

□ For sample preparation, ○ for the method, ■ for detection

CHEMICALS, CONTINUED

Method	1	2	3	4	5	6	7	8	9	10	11	12	13	14	15
Carbamylyte; standards for 2D PAGE						○		○		○					
CelloSeal®	○			○		○		○		○					○
Coomassie Brilliant Blue G-250		■	■			■	■		■	■					
Coomassie Brilliant Blue R-350		■	■			■	■		■	■					■
Cupric sulphate			■			■			■						
Dextran gel Sephadex IEF										○					
Dithiothreitol						□	□		□	○					□
EDTA-Na2		○□				□	□		□	○	○□	○□	○□	○□	□
Ethanol						■	■		■	■	■	■	■	■	■
Ethidium bromide															■
Ethylenglycol						○	○			○	○				○
Dyes for marking the front: Orange G, Bromophenol Blue	□			□			□			□					
Xylencyanol															
Pyronine, Basic Blue (cationic)				□○											
Formaldehyde solution 37%	■				■	■	■		■	■	■	■	■	■	■
Formamide												□	○	□	
Glutaraldehyde solution 25%						■	■		■	■					■

CHEMICALS, CONTINUED

Method	1	2	3	4	5	6	7	8	9	10	11	12	13	14	15
Glycerol	○						○		○	○	○	○	○	○	○
Glycine		■	■	■	■	■	■○		■	■○	■	■	■	■	■○
HEPES				○											
Hydrochloric acid, fuming							○		○	○					○
Immobiline II Starterkit, *assorted (6 × 10 mL)*									○	○					
Indian Ink								■							
Iodoacetamide						□	□			□					□
Kerosene or DC-200 silicon oil; *for cooling*	○	○	○	○	○	○	○	○	○	○	○	○	○	○	
Marker proteins for MW															
Calibration kit (14 400 to 94 000 Da)							○								○
Collagen solution up to 300 kDa							○								○
Peptide marker							○								○
Marker proteins for pI, depending on the pH interval					○				○						
2-Mercaptoethanol						□			○□	○□					□
Methanol	■		■	■	■	■	■	○	■	■					■
NAP columns for DNA purification, *to desalt protein*	□			□	□	□	□				□	□	□	□	□

CHEMICALS, CONTINUED

Method	1	2	3	4	5	6	7	8	9	10	11	12	13	14	15
Non-ionic detergents: Nonidet NP-40, Triton X-100, ProSolv II:				□○		□○			□○	□○					
Pharmalyte® carrier ampholyte, *depending on the pH interval*			○		○				□	□					
PhastGel®Blue R, *staining tablets Coomassie R-350*	■		■	■	■	■	■							■	
Phosphoric acid							■				○		○	○	
PMSF, *protease inhibitor*						□	□	□	□	□	○	○	○	○	□
Repel Silane	○		○	○	○	○	○	○	○	○	○	○	○	○	
SDS							□○	○	□○	□○	○	○			□○
Silver nitrate	■				■	■		■	■		■	■	■	■	■
Soda lime pellets *(CO2 trap)*					○	○		○	○						
Sodium acetate	■				■	■		■	■		■		■		■
Sodium carbonate				■		■			■		■	■	■	■	■
Sodium chloride	■							■							
Sodium dihydrogen phosphate	○							■							
Sodium hydrogen carbonate							■			■					
Sodium hydrogen phosphate	○							■							

CHEMICALS, CONTINUED

Method	1	2	3	4	5	6	7	8	9	10	11	12	13	14	15
Sodiumhydroxid					○	○		■	○	○		□			
Sodium thiosulphate							■			■	■	■	■	■	■
Sorbitol		○			○	○									
Sulfuric acid conc.			■			■			■						
Sulphosalicylic acid						■									
TEMED	○		○	○		○	○		○	○	○	○	○	○	○
TMPTMA								■							
Trichloracetic acid	■	■	■	■	■	■			■						
Tricine	○				○		○			○	○	○	○	○	○
TRIS	○					○	○	○	○	○	○	○	○	○	○
Tungstosilicic acid	■		■		■	■			■						
Tween 20								■							
Urea						□○			□○	□○	□	□	○	□○	

All chemicals must be of analytical (p.A.) quality.

□ For sample preparation, ○ for the method, ■ for detection

Method 1: PAGE of dyes

In most laboratories electrophoresis is used for the separation of relatively high molecular weight substances, such as proteins and nucleic acids. Substances with low molecular weights such as polyphenols and dyes are separated by column or thin-layer chromatography.

High molecular weight:
> 10 kDa
Low molecular weight: < 1 kDa

In the following part, simple electrophoretic methods for the separation of substances of low molecular weight will be described using dyes as an example.

Foodstuffs, car paints, cosmetic dyes etc.

1 Sample preparation

Ten milligrams of each dye are dissolved in 5 mL of distilled water, 1.5 µL is applied for each run.

2 Stock solutions

Acrylamide, Bis (T= 30%, C=3%)

29.1 g of acrylamide + 0.9 g of Bis, made up to 100 mL with double-distilled water

Dispose of the remains ecologically: polymerize with an excess of APS.

Caution! *Acrylamide and Bis are toxic in the monomeric form. Avoid skin contact and do not pipette by mouth.*

The solution is stable (for IEF) for one week when stored in the dark at 4 °C (refrigerator):

but for this method, the solution can be kept for several weeks.

Ammonium persulfate solution (APS) 40% (w/v):

dissolve 400 mg of ammonium persulfate in 1 mL of double-distilled water.

stable for one week when stored in the dark at 4 °C.

0.75 mol/L phosphate buffer pH 7.0:

38.6 g of Na_2HPO_4
8.25 g of NaH_2PO_4
fill up to 500 mL with distilled water

3 Preparing the casting cassette

This method works best with very thin gels (0.25 mm),

● because high fields can be applied

fast separation → limited diffusion

● because the substances which are separated can be fixed by rapid drying.

chemical fixing is not possible.

Gasket

Two layers of Parafilm® (50 cm wide) are superimposed and cut with a knife so that a U-shaped gasket is formed and adheres to the glass plate.

A scalpel is too sharp. The knife presses the edges together so the layers stay together.

Slotformer

Sample application is done in small sample wells which are polymerized in the surface of the gel. To form the sample wells in the gel a template must be fixed to the glass plate. A cleaned and degreased glass plate is
- placed on the template pattern (in the appendix)
- fixed to the work surface.

Place two layers of "Scotch tape" (one layer 50 µm) on the starting point and smooth them down so that no bubbles appear. Cut the slot former out with a scalpel to 1×7 mm (Fig. 1). After pressing out the holes of the slot former, remove the remains of the cellophane with methanol.

The sample wells are placed in the middle since there are anionic and cationic dyes.

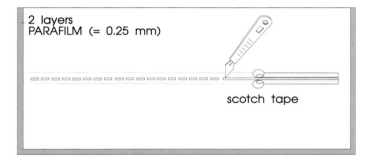

Fig. 1: Preparing the slot former.

If longer separation distances are desired, the slot former is placed closer to the edge. Anionic substances are applied near the cathode and cationic ones near the anode.

The slot former is then made hydrophobic. A few mL of Repel Silane are spread evenly over the whole slot former with a tissue under the fume hood. When the Repel Silane is dry, the chloride ions which result from the coating are rinsed off with water.

This operation only needs to be carried out once.

The Parafilm® gasket is placed along the edge of the slot former plate (Fig. 1).

When one side of the gasket is coated with CelloSeal®, it adheres to the glass plate.

Assembling the gel cassette

The gel is covalently polymerized on a plastic film for mechanical support and easier handling. Place the glass plate on a clean absorbent paper towel and moisten it with a small volume of water. Apply the GelBond PAG film on the glass plate with a rubber roller, placing the untreated hydrophobic side down (Fig. 2). A thin layer of water then forms between the glass plate and the film and holds them together by adhesion. The excess water which runs out the sides is soaked up by the tissue.

To facilitate pouring the gel solution, the film should overlap the length of the glass plate by 1 mm.

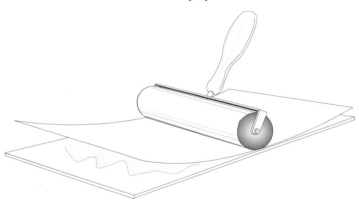

Fig. 2: Applying the support film with a roller.

Place the slot former over the glass plate and clamp the cassette together (Fig. 3).

Slot templates facing down.

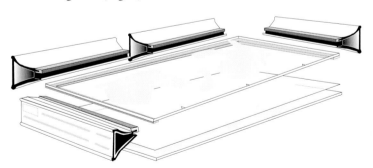

Fig. 3: Assembling the gel cassette.

Since a very thin layer of gel is needed for this method, the cassette is wedged open with paper clips and the clamps are pushed low down on the sides (Fig. 4).

This prevents the formation of air bubbles in the layer.

4 Casting ultrathin-layer gels

Recipe for two gels

0.33 mol/L phosphate buffer, pH 7.0 (T = 8%, C = 3%)

Mix in test tubes with screw caps (15 mL):
4.0 mL of acrylamide, Bis solution
2.5 mL of glycerol (87%)
4.0 mL of phosphate buffer
fill up to 15 mL with double-distilled water
7 µL of TEMED (100%)
15 µL of APS

Immediately after mixing pour in 7.5 mL per gel using a pipette or 20 mL syringe (Fig. 4).

A higher T value leads to less sharp bands.

Glycerol has two functions:
● it prevents diffusion when proteins are applied
● it keeps the gel elastic during drying.

100 µL of 60 % (v/v) isopropanol-water are then layered on the edge of the solution. Isopropanol prevents oxygen, which inhibits polymerization, from diffusing into the gel. The gel will then present a well-defined, aesthetic upper edge.

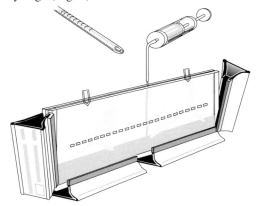

Fig 4: Casting an ultrathin-layer polyacrylamide gel: after the calculated amount of polymerization solution has been added, remove the paper clips and push the clamps back into position.

● Let the gel stand for one hour at room temperature.

Polymerization

5 Electrophoretic separation

● Switch on the cooling system: +10 °C.

Removing the gel

● Remove the clamps and lay the sandwich on the cooling plate at 10 °C with the glass plate containing the slot former on the bottom.

Cooling gives the gel a better consistency and it usually begins to separate from the slot former in the cassette already.

● Hold the cassette vertically and lift the GelBond film from the glass plate with a thin spatula.

● Pull the film and the gel from the slot former.

● Coat the cooling plate with 2 mL of the *contact fluids* kerosene or DC-200 silicone oil.

Water and other liquids are inadequate.

● Place the gel on the cooling plate with the film on the bottom. Avoid air bubbles.

● Lay two of the electrode wicks into the compartements of the PaperPool (if smaller gel portions are used, cut them to size). Apply 20 mL of phosphate buffer to each wick (Fig. 5). Place one strip onto the anodal edge of the gel and the other strip onto the cathodal edge, each overlapping the gel edges by 5 mm (Fig. 6).

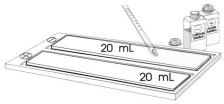

Fig. 5 : Soaking the wicks in electrode buffer in the PaperPool.

Smooth out air bubbles by sliding bent tip foreceps along the edges of the wicks laying in contact with the gel.

● Quickly and carefully pipette the sample into the sample wells: 1.5 µL Apply pure colors and mixtures of dyes but keep acid and basic dyes separate.

Acid and basic dyes precipitate when combined!

● Clean platinum electrode wires before (and after) each electrophoresis with a wet tissue paper.

● Move electrodes so that they will rest on the outer edge of the electrode wicks. Connect the cables of the electrodes to the apparatus and lower the electrode holder plate (Fig. 6).

● Close the safety lid.

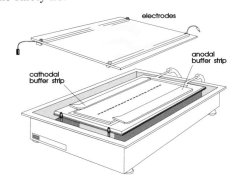

Fig. 6: Gel and electrode wicks for the separation of dyes. The samples are applied in the middle.

Running conditions:
Power supply: 400 V_{max}, 60 mA $_{max}$, 20 W_{max}, about 1 h.

After separation

● Switch off the power supply,

● Open the safety lid,

● Remove the electrode wicks,

● Remove the gel and dry immediately.   *It is best to dry the gel on a warm surface, for example on a light box (switched on!).*

   The separation of different dyes is shown in Fig. 7. Several food colorings can be seen to consist of mixtures.

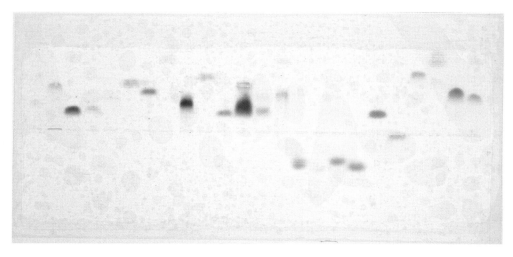

Fig. 7: Ultrathin-layer electrophoresis of dyes. Anode at the top.

Method 2: Agarose and immunoelectrophoresis

The principles and uses of agarose gel, affinity and immuno-electrophoresis are described in part I.

see pages 13 and following

Because of the large pores, agarose electrophoresis is especially suited to the separation of lipoproteins and immunoglobulins, and to specific detection by immunofixation.

Immunoelectrophoresis according to Grabar and Williams (1953) and Laurell (1966) is not only used in clinical diagnostic and pharmaceutical production, it is also an official method for detection of falsifications and the use of forbidden additives in the food industry.

Immunodiffusion is another possible method:

Ouchterlony Ö. Allergy. 6 (1958) 6.

These methods are traditionally carried out in a Tris-barbituric acid buffer (veronal buffer). A few years ago the use of barbituric acid was limited by the drug law (Susann, 1966). For this reason the following methods will use a Tris-Tricine buffer.

Susann J. The Valley of the Dolls. Corgi Publ. London (1966).

1 Sample preparation

● Marker proteins pI 5.5 to 10.7 + 100 μL of double-distilled water.

Apply 6.5 μL

● Store deep-frozen portions of meat extract from pork, rabbit, veal, beef. Dilute before use: 100 μL of meat extract + 300 μL of double-distilled water.

Apply 6.5 μL.

● Other samples:

Set the protein concentration around 1 to 3 mg/mL. Dilute with double-distilled water. The salt concentration should not exceed 50 mmol/L.

Apply 6.5 μL

Desalting with a NAP-10 column may be necessary: apply 1 mL of sample – use 1.5 mL of eluent.

2 Stock solutions

Tris-Tricine lactate buffer pH 8.6:

117.6 g of Tris
51.6 g of Tricine
12.7 g of calcium lactate
2.5 g of NaN$_3$
make up to 3 L with distilled water.

For agarose electrophoresis an amphoteric buffer system can be used, for example use 0.6 mol/L HEPES as in method 4. Agarose is mixed with the rehydration solution, boiled and poured on the gel.

Physiological salt solution (0.15 mol/L):
  9 g of NaCl, make up to 1 L with distilled water.

Bromophenol Blue solution:
  10 mg, make up to 10 mL with distilled water.

Agarose gel solution:
  1 g of Agarose L
  100 mL of Tris-Tricine lactate buffer
● Sprinkle the dry agarose on the surface of the buffer solution (to prevent the formation of lumps) and heat in a microwave oven on the lowest setting till the agarose is melted.
● Pour the solution in test tubes (15 mL each)

  The test tubes can be kept in the refrigerator at 4 °C for several months.

3 Preparing the gels

a) Agarose gel electrophoresis
  Agarose gel electrophoresis also works best when the samples are pipetted into small wells. To form these sample wells in the gel, a mould must be fixed on to the spacer glass plate.

The "spacer" is the glass plate with the 0.5 mm thick, U-shaped silicone rubber gasket.

Preparing the slot former

  The cleaned and degreased glass plate with the 0.5 mm U-shaped silicone rubber gasket is placed on the slot former template (slot former template in the appendix). On the place which will be used as starting point, a layer of "Dymo" tape is applied (embossing tape, 250 μm thick) on the glass plate avoiding air bubbles. Cut out the slot former with a scalpel (Fig. 1). After pressing the slots once more against the glass plate, remove the remains of plastic with methanol.

"Dymo" tape with a smooth adhesive surface should be used. When a structured adhesive surface is used, small air bubbles can be enclosed and holes appear around the wells.

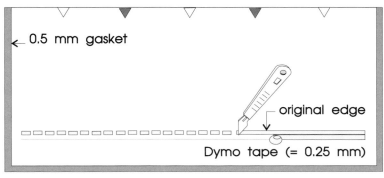

Fig. 1: Preparing the slot former.

The mold is then made hydrophobic. A few mL of Repel Silane are spread over the whole slot former with a tissue under the fume hood. When the Repel Silane is dry, the chloride ions resulting from the coating are washed off with water.

This operation only needs to be carried out once.

Assembling the gel cassette

The gel is molded on a support film for better mechanical stability and easier handling. Place a glass plate on an absorbent tissue and moisten it with a few mL of water. Apply the GelBond film with a rubber roller placing the untreated hydrophobic side on the bottom (Fig. 2). A thin layer of water then forms between the glass plate and the film and holds them together by adhesion. The excess water which runs out is soaked up by the tissue. To facilitate the pouring of the gel solution, the film should overlap the length of the glass plate by about 1 mm.

GelBond film is used for agarose: it is a polyester film with a coating of dry agarose.

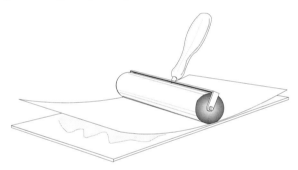

Fig. 2: Applying the support film with a roller.

The slot former is placed over the glass plate and the cassette is clamped together (Fig. 3). Before filling with the hot agarose solution, the cassette and a 10 mL glass pipette should be warmed in a heating cabinet at 75 °C.

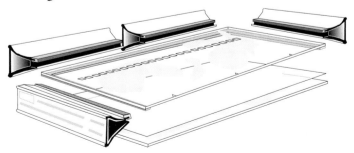

Fig. 3: Assembling the gel cassette.

● Take a 15 mL test tube out of the refrigerator and liquefy the contents in the microwave.

● Remove the cassette from the heating cabinet; draw the hot agarose solution in the pipette with a rubber bulb and quickly release the solution in the cassette (Fig. 4).

Avoid air bubbles; should some appear nevertheless, remove them with a long strip of polyester film.

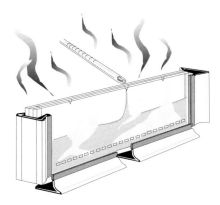

Fig. 4: Pouring the hot agarose solution into the prewarmed cassette.

● Let the cassette stand for 1 or 2 h at room temperature.

This allows the gel to set slowly.

● Remove the clamps and take out the gel.

● Place the gel on a wet filter paper and leave it overnight in a humidity chamber (Fig. 5) in the refrigerator. It can be kept up to one week under these conditions.

Only then does the definite agarose gel structure form (see page 10).

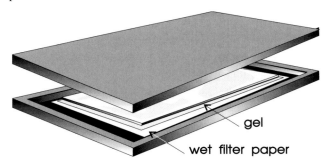

gel

wet filter paper

Fig. 5: Storing the agarose gel overnight in a humidity chamber.

b) Immunoelectrophoresis gels

Small gels are usually prepared, especially for the "rocket" technique, so as to be sparing with expensive antibodies.

In this technique, the antibodies are poured directly in the gel.

The *Grabar-Williams technique* as well will be described for small gels here.

Figure 11 gives an example in a large gel.

For these techniques it is advantageous to use the immuno-electrophoresis kit because it contains templates and foils in the right sizes for the holes and troughs.

● Cut out 8.4 cm wide pieces of GelBond film.

They can also be bought in this size.

To ensure that the gel thickness is uniform, the solution is poured on the film on a leveling table. 12 mL of gel solution are used for each plate.

● Use a spirit level to ensure that the levelling table is horizontal.
● Place a sheet of GelBond with the hydrophilic side up in the middle of the table.
● Liquefy the gel solution (in the test tube) by warming it.

When antibodies have to be added to the gel ("rocket" technique): cool the gel to 55 °C, add about 120 μL of antibody solution*, mix the solution (avoid air bubbles).

* This is an indicative value, it depends on the antigen-antibody titer.

● Pour the gel solution on the GelBond film and let the gel set.

The gels can be stored in a humidity chamber in the refrigerator.

Punching out the sample wells and troughs

The gel puncher is connected to a water jet pump with thin vacuum tubing so that the pieces of gel can be sucked away immediately.

Choose the size of the wells according to the sample concentration and antibody titer.

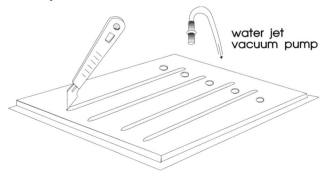

water jet
vacuum pump

Fig. 6: Punching out the sample wells and troughs.

Grabar-Williams gel (Fig. 6)

● Place the gel (without antibodies) on the separation bed.

● Use the Grabar-Williams template.

● Punch out 5 sample wells at places which will later be near the cathodic side   *The other holes in the template are for agar gels.*

● Excise the troughs with a scalpel; the gel strips remain in the gel during electrophoresis.

"Rocket" gel (Fig. 7):

● Place the gel (with antibodies) on the separation bed.

● Use the Laurell template.

● Punch out 8 wells: use every second hole.

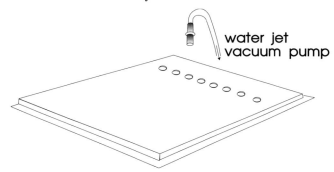

water jet vacuum pump

Fig. 7: Punching the sample wells for rocket immunoelectrophoresis.

4 Electrophoresis

● Switch on the cooling system (10 °C).

● Place the cooling plate on the side, beside the Multiphor.

● Place the electrodes (orange plastic plates) in the inner compartment of the tanks and plug them in.

● Pour in the Tris-Tricine lactate buffer, 1 L per tank; put the cooling plate back in.

● Coat the cooling plate with 1 mL of contact fluid, either kerosene or DC-200 silicone oil.   *Do not use water, as it can cause electrical shorting.*

● Place the gel with the film on the bottom on the cooling plate (Fig. 6).

● Dry the surface with filter paper (Fig. 8) because agarose gels have a liquid film on the surface.   *This is also true for immunoelectrophoresis gels.*

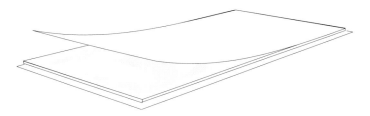

Fig. 8: Drying the surface of the gel with filter paper.

● Place the gel on the cooling plate, the sample wells should be on the cathodal side (Fig. 9).

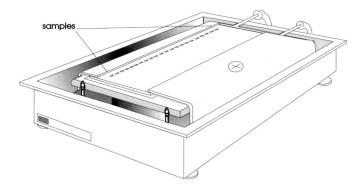

Fig. 9: Agarose gel electrophoresis.

● Soak 8 layers of electrode wicks or 1 layer of Wettex permanent electrode strip in the buffer and place them on the gel to ensure good contact between the gel and the buffer.

The wicks should overlap the gel by about 1 cm.

● Rapidly pipette 5 mL of each sample in the trough.

The samples should not diffuse.

● Close the safety lid.

● Switch on the power supply.
Running conditions: maximum 400 V, 30 mA, 30 W.

● Fixing, staining etc, should be done directly after the run.

Grabar-Williams technique

In general sample application follows the scheme: sample – positive control – sample – positive control – sample.

● Pipette the samples and controls in the troughs (5 to 20 mL)

● Add one drop of Bromophenol Blue solution and start the run.

Separation conditions:

Field strength 10 V/cm (Check with a voltage probe).
Temperature 10 °C.
Time about 45 min (till Bromophenol Blue has reached the edge).

● Place the gel in the gel holder and with the "shovel" on the back of the scalpel remove the gel strips from the troughs.

● Place the gel in the humidity chamber.

● Pipette 100 μL of antibody in each trough.

● Let the solution diffuse for about 15 h at room temperature.   *The precipitin arcs then form.*

● Staining is only done after the non-precipitated antigens and antibodies have been washed out (see below).   *The precipitates remain in the gel.*

Laurell technique

● Pipette the samples in the wells of the antibody containing gel as quickly as possible to prevent diffusion.

In general a series of dilutions (4 samples) of an antibody solution is run at the same time so as to obtain a concentration calibration curve.

● Start the electrophoresis immediately after sample application.

Separation conditions:

Field strength	10 V/cm (Check with voltage probe)
Temperature	10 °C
Time	3 h

● Staining should be done after the non-precipitated antigens and antibodies have been washed out (see below).

The buffer is set at pH 8.6 where the specific antibodies in the gel have a minimal net charge and no electrophoretic mobility (Caution! Antibodies can have different optimum pH values depending on their origin; for rabbit antibodies pH 7.8 is usually used).   *The sample proteins (antigens) are charged at pH 8.6 and migrate in the gel.*

The determination of antigens with a very low or no net charge at this pH is problematic (e.g. IgG). However, the isoelectric point of these proteins can be lowered by acetylation or carbamylation and their electrophoretic mobility thus increased.

At first the antigen molecules are in excess, so the antigen-antibody complexes are soluble and migrate towards the anode. The equivalence concentration is reached along both sides of the migrating antigen track and a rocket-shaped precipitin line is formed continuously starting at the bottom.   *It is important that the antibodies be of good quality, otherwise the lines are not well defined or several "rockets" are formed in another.*

The area of the "rocket" is directly proportional to the antigen concentration. Exact concentration determination is done by measuring the area of the "rocket". But in many cases it is sufficient to measure the height of the rocket.

5 Protein detection

Coomassie staining (Agarose electrophoresis)

Solution with H_2O_{dist}

- *Fixing:* 30 min in 20% (w/v) TCA;

- *Washing:* 2×15 min in 200 mL of fresh solution each time: 10% glacial acetic acid, 25% methanol;

- *Drying:* cover the gel with 3 layers of filter paper and place a glass plate and a weight (1 to 2 kg) over them (Fig. 10). Remove after 10 min and finish drying in a heating cabinet;

  First humidify the filter paper that lies directly on the agarose.

- *Staining:* 10 min in 0.5% (w/v) Coomassie R-350 in 10% glacial acetic acid, 25% methanol: dissolve 3 tablets of PhastGel Blue (0.4 g dye each) in 250 mL;

- *Destaining:* in 10% glacial acetic acid, 25% methanol till the background is clear:

- *Drying:* in the heating cabinet.

Immunofixing of agarose electrophoresis

With this fixing method only the protein bands which have formed an insoluble immunoprecipitate with the antibody are fixed.

All the other proteins and the excess antibody are washed out of the gel with a NaCl solution.

Antibody solution: 1:2 (or 1:3 depending on the antibody titer) dilute with double-distilled water and apply on the surface of the gel with a glass rod or a pipette: about 400 to 600 mL;

- *Incubation:* 90 min in the humidity chamber in an incubator or a heating cabinet at 37 °C;

- *Pressing:* 20 min with 3 layers of filter paper, a glass plate and a weight (Fig. 10);

  First humidify the filter paper that lies directly on the agarose.

- *Washing:* in physiological sodium chloride solution (0.9% NaCl w/v) overnight;

- *Drying:* see Coomassie staining;

- *Staining* and destaining: as for Coomassie staining.

It is important to know the antibody titer, since for example, hollow bands can occur when the antibody solution is too concentrated: in the middle one antibody binds to one antigen and no precipitate results.

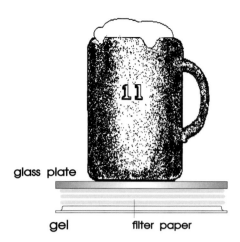

glass plate

gel   **filter paper**

Fig. 10: Pressing the agarose gel.

Coomassie staining (immunoelectrophoresis)

In *"rocket"* electrophoresis as in *Grabar-Williams* electro-phoresis, the gel contains antibodies which must be washed out before staining.

● *Pressing:* 20 min with 3 layers of filter paper, a glass plate   *First humidify the filter paper*
  and weight (Fig. 10);   *that lies directly on the gel.*

● *Washing:* in physiological sodium chloride solution (0.9% NaCl w/v) overnight;

● *Drying:* see Coomassie staining;

● *Staining* and destaining: as for Coomassie staining.

Silver staining

If the sensitivity of Coomassie staining is not sufficient, silver   *Willoughby EW, Lambert A.*
staining can be performed on the dry gel (Willoughby and   *Anal Biochem. 130 (1983) 353-*
Lambert, 1983):   *358.*

Solution A: 25 g of Na_2CO_3, 500 mL of double-distilled water;

Solution B: 1.0 g of NH_4NO_3, 1.0 g of $AgNO_3$, 5.0 g of tungstosilisic acid, 7.0 mL of formaldehyde solution (37%), make up to 500 mL with double distilled water.

Staining: Mix 35 mL of solution A and 65 mL of solution B just before use, put the gel in the resulting whitish solution and incubate while agitating until the desired intensity is reached. Briefly rinse with double-distilled water;

Stop with 0.05 mol/L glycine. Remove the remains of metallic silver from the gel and the support film with a cotton swab;

Air dry.

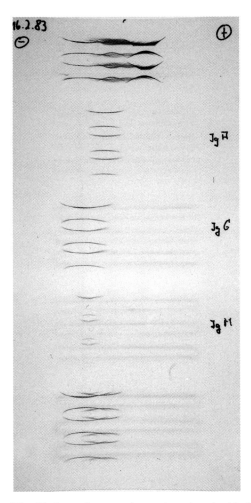

Fig. 11: Large immunoelectrophoresis gel with the Grabar-Williams technique.

Method 3: Titration curve analysis

The principles of titration curve analysis are described in part 1. *see page 57*

   The technical execution with washed and rehydrated polyacrylamide gels will be explained here. This variant excludes the influences of ionic catalysts which are found in the gel after polymerization.

   When it is not necessary to obtain very precise results, the carrier ampholyte can be polymerized directly into the gel.

APS and TEMED can influence the formation of the pH gradient.

This saves washing, drying and rehydration.

1 Sample preparation

● Marker proteins pI 4.7 to 10.6 or

● Marker proteins pI 5.5 to 10.7 + 100 µL of double distilled water.

Apply 50 µL

● Meat extract from pork, rabbit, veal, beef, store in frozen portions. Dilute before use: 100 µL of meat extract + 300 µL of double-distilled water.

Apply 50 µL.

Other samples:
   Set the protein concentration around 1 to 3 mg/mL.
   Dilute with double-distilled water. The salt concentration should not exceed 50 mmol/L: apply 1 mL of sample solution - use 1.5 mL of eluent.

Apply 50 µL

2 Stock solutions

Acrylamide, Bis (T = 30%, C = 3%)
29.1 g of acrylamide + 0.9 g of Bis, make up to 100 mL with double-distilled water or
   reconstitute PrePAG Mix (29.1:0.9) with 100 mL of double-distilled water.

Dispose of the remains ecologically, polymerize with an excess of APS.

   Caution! *Acrylamide and Bis are toxic in the monomeric form. Avoid contact with the skin, do not pipette by hand.*

Can be stored for one week in a dark place at 4 °C (refrigerator).

Can still be used for SDS gels for several weeks.

Ammonium persulfate solution (APS) 40% (w/v):
   Dissolve 400 mg of ammonium persulfate in 1 mL of double-distilled water.

Can be stored for one week in the refrigerator (4 °C).

0.25 mol/L Tris-HCl buffer :

3.03 g of Tris + 80 mL of double-distilled water, titrate to pH 8.4 with 4 mol/L HCl and make up to 100 mL with double-distilled water.

This buffer corresponds to the anode buffer of horizontal electrophoresis as in method 7. It is used to set the pH value of the polymerization solution, since this works best in a slightly basic medium.

The buffer is removed before washing.

Glycerol 87% (w/v):

Glycerol fulfills several roles: in the polymerization solution it improves the fluidity and makes the solution denser at the same time so that overlayering is easier. In the last washing solution, glycerol ensures that the gel does not roll up upon drying.

3 Preparing the blank gels

Preparing the casting cassette

In titration curve analysis the sample is applied in long narrow troughs which run along the whole pH gradient which is electrophoretically established beforehand. To form the sample troughs a mould must be fixed on a glass plate. Since square gels are used for this method, two gels are poured together with the glass plate of the gel kit and rehydrated later.

The gel is later cut in half so two separations can be run simultaneously. If only one gel is needed, the second half can be stored in the humidity chamber in the refrigerator (2 weeks at most).

Two 10 cm long pieces of "Dymo" tape (embossing tape, 250 μm thick) are cut and applied on the cleaned and degreased glass plate with a 0.5 mm thick U-shaped gasket. As shown in Fig. 1, a piece is cut away with the scalpel in such a way that 1 to 2 cm Dymo tape remain. After pressing the tape once more against the glass plate, the remains of the tape are removed with methanol.

Dymo tape with a smooth adhesive surface should be used. When the adhesive surface is structured, small air bubbles which inhibit polymerization can be enclosed.

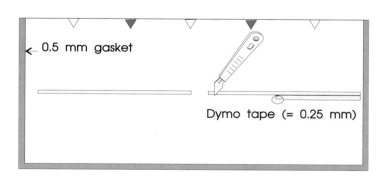

Fig. 1: Cutting the narrow adhesive tape to form the sample troughs in the gel.

When only one gel is made and rehydrated, the rehydrating cassette can be used as a mold. In this case, a piece of tape is stuck on the bottom half of the glass plate with the gasket.

see methods 7, 10 and 11

This mold is then made hydrophobic. A few mL of Repel Silane are spread over the whole slot former with a tissue under the fume hood. When the Repel Silane is dry, the chloride ions which result from the coating are rinsed off with water.

This operation only needs to be carried out once.

Assembling the gel cassette

● Remove the GelBond PAG film from the package.

The hydrophobic side can be identified with a few drops of water.

Pour a few mL of water on the glass plate and place the support film on it with the hydrophobic side down. Press the support film onto the glass plate with a roller (Fig. 2).

This facilitates filling the mold later on.

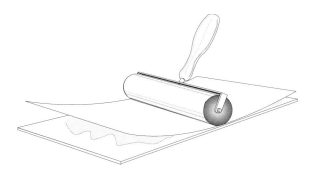

Fig. 2: Rolling on the support film.

The spacer is then placed on the glass plate with the gasket facing downwards and the cassette is clamped together (Fig. 3).

The "spacer" is the glass plate with the 0.5 mm thick U-shaped silicone rubber gasket.

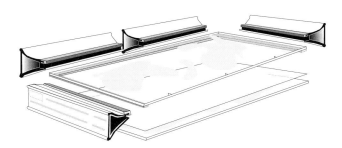

Fig. 3: Assembling the gel cassette.

● *Composition of the polymerization solution for a gel with*
  T = 4.2%.

2.7 mL of acrylamide/Bis
0.5 mL of Tris-HCl
1 mL of glycerol
10 μL of TEMED
make up to 20 mL with double-distilled water.

If the gel is used right away (without washing), 2 g of sorbitol and 1.3 mL of Ampholine or Pharmalyte should be used instead of glycerol and Tris-HCl.

Filling the gel cassette

When everything else is ready, 20 μL of APS is added to the polymerization solution.

Add 40 μL of APS for cast gels (no washing)!

The cassette is filled with a 10 mL pipette or a 20 mL syringe (Fig. 4). Draw the solution into the pipette (a 10 mL scale plus the headspace make 18 mL) and place it in the middle notch. Slowly release the solution it will be directed into the cassette by the strip of film sticking out.

Never pipette the toxic monomer solutions by mouth!

When the rehydration cassette is used, introduce the solution from below with a syringe.

Introduce 100 μL of 60 % v/v isopropanol-water into each notch. Isopropanol prevents oxygen, which inhibits polymerization, from diffusing into the gel. The gel will then have a well defined aesthetic edge.

Air bubbles which might occur can be dislodged out with a long strip of polyester film.

Let the gel stand for one hour at room temperature.

Polymerization

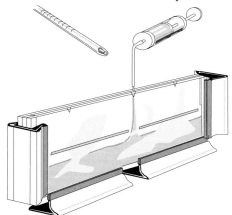

Fig. 4: Pouring the polymerization solution.

Removing the gel from the cassette

After the gel has polymerized for one hour remove the clamps and lift the glass plate off the film with a spatula. Peel the gel of the spacer by slowly pulling on a corner of the film.

Washing the gel

Wash the gel in double-distilled water by shaking it three time for 20 min. Add 2% glycerol to the last washing solution. If a MultiWash appliance is used: deionize for about 30 min by pumping double-distilled water through the mixed bed ion-exchange cartridge (Fig. 5).

This washes the remains of monomers, APS and TEMED out of the gel.

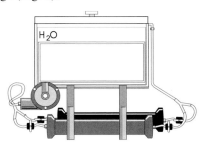

Fig. 5: MultiWash. Gel staining and washing apparatus.

Drying the gel

Dry the gel overnight at room temperature. Then cover the gels with a protecting film and store them frozen (−20 °C) in sealed plastic bags.

Heat drying damages the swelling capacity, which is why the gels should be stored frozen.

4 Titration curve analysis

Reswelling the gel

Place the GelPool on a horizontal table. Select the appropriate reswelling chamber of the GelPool. Clean it with distilled water and tissue paper. Pipet the *rehydration solution* (one gel):

        400 mL ethylenglycol (= 7.5% v/v)
        390 µL of Ampholine
        or Pharmalyte (= 3% w/v)
        make up to 5.2 mL with double-distilled water.

Ethylenglycol increases the viscosity of the solution, so the curves become smoother.

Set the edge of the gel-film − with the dry gel surface facing downward − into the rehydration solution (fig. 6 A) and slowly lower the film down. At the same time move the gel-film to and fro, in order to achieve an even distribution of the liqid and to avoid trapping airbubbles.

Lift the film at the edges with foreceps, and slowly lower them down, in order to maintain an even distribution of the liquid (fig. 6 B) and to remove airbubbles.

Note: Repeat this measure several times during the first 15 min, to prevent the gel from sticking to the GelPool.

Check, whether the gel can be moved around on its reswelling liquid.

60 min later the gel has reswollen completely and can be removed from the GelPool.

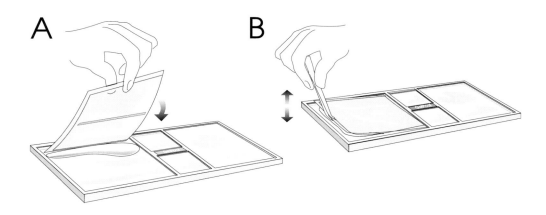

Fig. 6: Rehydration of a titration curve gel in the GelPool.

● Dry the sample trenches with filter paper.

a) Formation of the pH gradient

Run without sample

● Switch on the cooling system: +10 °C.

● Pipette 1 mL of contact fluid, kerosene or DC-200 silicone oil, on to the middle of the cooling plate.

Do not use water! This can cause electrical shorting.

● Place the gel in the middle of the gel with the support film on the bottom. The sample troughs should lie perpendicular to the electrodes (Fig. 7).

This method must be carried out at a defined temperature since the pH gradient and the net charge are temperature dependent.

● The power supply settings are the following:

If two gels are run in parallel the current and power must be doubled.

1500 V_{max}, 7 mA_{max}, 7 W_{max}.

The electrodes are placed directly on the edge of the gel (Fig. 7).

Electrode strips are not needed for washed gels.

● Plug in the cable. Take care that the long anode cable is hooked to the front.

● Close the safety lid.

● Switch on the power supply.

A continuous pH gradient from 3.5 to 9.5 will form in about 60 min.

see chapter 3.

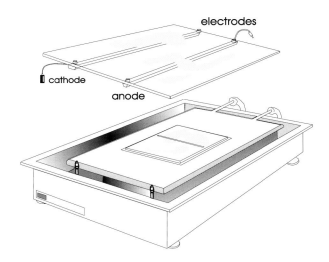

electrodes

cathode

anode

Fig. 7: Placing the titration gel on the cooling plate for the formation of the pH gradient.

b) Native electrophoresis in the pH spectrum

● Once the power is switched off, the safety lid opened and the electrodes are removed, the gel is turned by 90°: The alkaline side of the gel is now directed towards the side of the cooling tubing!

This standard is used when placing the gels and presenting results to simplify the interpretation and comparison of results.

This means that the proteins in the left side of the gel are positively charged and migrate towards the cathode while those in the right side are negatively charged and migrate towards the anode (Fig. 9 and 36B, Chapter 3.7).

● Using a pipette introduce 50 μL of sample solution (1 mg/mL) into the trough.

Work quickly to prevent diffusion from destroying the gradient.

● Proceed with the electrophoresis immediately: 1200 V_{max}, 10 mA_{max}, 5 W_{max}.

The proteins with different mobilities now migrate towards the anode or the cathode with velocities which depend on the actual pH of the medium.

Titration curves are formed.

● Stop after 20 min and stain the gel.

5 Coomassie and silver staining

Colloidal Coomassie staining

The result is quite quickly visible with this method. Few steps are necessary, the staining solutions are stable and there is no background staining. Oligopeptides (10 to 15 amino acids) which are not properly fixed by other methods can be revealed here. In addition, the solution is almost odorless (Diezel *et al.* 1972; Blakesley and Boezi, 1977).

Diezel W, Kopperschläger G, Hofmann E. Anal Biochem. 48 (1972) 617-620.
Blakesley RW, Boezi JA. Anal Biochem. 82 (1977) 580-582.

Preparation of the staining solution:

Dissolve 2 g of Coomassie G-250 in 1 L of distilled water and add 1 L of sulfuric acid (1 mol/L or 55.5 mL of conc H_2SO_4 per L) while stirring. After stirring for 3 h, filter (paper filter) and add 220 mL of sodium hydroxide (10 mol/L or 88 g in 220 mL). Finally add 310 mL of 100% TCA (w/v) and mix well, the solution will turn green.

Fixing and staining: 3 h at 50 °C or overnight at room temperature in the colloidal solution;

Washing out the acid: soak in water for 1 or 2 h, the green color of the curves will become blue and more intense.

Fast Coomassie staining

Stock solutions:

Use distilled water for all solutions

TCA: 100% TCA (w/v) 1 L
A: 0.2% (w/v) $CuSO_4$ + 20% glacial acetic acid
B: 60% methanol
C: dissolve 1 Phast Blue R tablet
in 400 mL of double distilled water,
add 600 mL of methanol
and stir for 5 to 10 min.

1 tablet = 0.4 g of Coomassie Brilliant Blue R-350

Staining:

● *Fixing:* 10 min in 200 mL of 20% TCA;

● *Washing:* 2 min in 200 mL of washing solution (mix equal parts of A and B);

● *Staining:* 15 min in 200 mL of 0.02% (w/v) R-350 solution at 50 °C while stirring (mix equal parts of A and C) (Fig. 8);

● *Destaining:* 15 to 20 min in washing solution at 50 °C while stirring;

● *Preserving:* 10 min in 200 mL of 5% glycerol;

● *Drying:* air-dry.

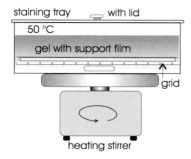

Fig. 8: Appliance for hot staining.

5 minute silver staining of dried gels

This method can be applied to stain dried gels already stained with Coomassie (the background must be completely clear) to increase sensitivity, or else unstained gels can be stained directly after pretreatment (Krause and Elbertzhagen, 1987). A significant advantage of this method is that no proteins or peptides are lost during the procedure. They often diffuse out of the gel during other silver staining methods because they cannot be irreversibly fixed in the focusing gels.

Krause I, Elbertzhagen H. In: Radola BJ, Ed.. Elektrophorese-Forum'87. This edition. (1987) 382-384.

according to the silver staining method for agarose gels (Willoughby and Lambert, 1983).

Pretreatment of the unstained gels:

● fix for 30 min in 20% TCA,

● wash for 2 ×5 min in 45% methanol. 10% glacial acetic acid,

● wash for 4 ×2 min in distilled water,

● impregnate for 2 min in 0.75% glycerol,

● air-dry.

Solution A: 25 g of Na_2CO_3, 500 mL of double distilled water;
Solution B: 1.0 g of NH_4NO_3, 1.0 g of $AgNO_3$, 5.0 g of tungstosilisic acid, 7.0 mL of formaldehyde solution (37%), make up to 500 mL with double-distilled water.

Silver staining:

Mix 35 mL of solution A with 65 mL of solution B just before use. Immediately soak the gel in the resulting whitish suspension and incubate while agitating until the desired intensity is reached. Briefly rinse with distilled water.

Stop with 0.05 mol/L glycine. Remove remains of metallic silver from the gel and support film with a cotton swab.

Air-dry.

6 Interpreting the curves

The schematic curves of three proteins A, B and C are shown in Fig. 9.

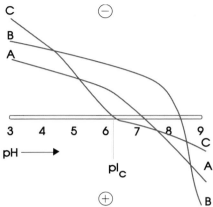

Fig. 9: Schematic representation of titration curves.

These results can be interpreted as follows for:

Ion-exchange chromatography

●   Flat curves point to bad separation in ion-exchangers, since a change in the pH in the buffer only has a limited influence of the net charge.

● The separation capacity is best when elution is carried out at a pH at which the curves lie far apart on the y-axis: in this example at pH 3.5 for a cation-exchanger. If an anion-exchanger is used and the elution is done at pH 7.7, protein B will not bind and will come out in the exclusion volume.

The intensity of the net charge and thus the binding to the separation medium present most differences there.

● If the intersections with the x-axis (pI) lie far apart, chromatofocusing will be effective.

A mixture of amphoteric buffers is used for elution in this case and the sample is fractionated according to the pIs.

Zone electrophoresis

● If the buffer with a pH value at which the curves intersect is chosen, these proteins can only be separated on the basis of their frictional resistance in a restrictive medium.

It is better to choose another pH value.

● For electrophoresis in a support free medium or for separations in non-restrictive media, choose a pH value at which the curves lie clearly apart on the y-axis: pH 5.3 or 7.7 in this example.

The mobilities are very different there.

Isoelectric focusing

● The protein C intersects the x-axis at a flat angle, this means that the mobility gradient is low at the pI. High field strength is necessary to focuse C.

In addition the focusing time must be increased.

● The curve of protein C lies along the x-axis above the pI. For IEF the sample must be applied at a pH < pI, otherwise C will migrate very slowly or not at all.

The protein only has a low charge here.

● If one or more proteins do not migrate out of the troughs or produce a smeared zone in a certain area, this means that the protein is unstable at this pH or that a few proteins aggregate (not shown).

The sample should not be applied in this interval.

Method 4: Native PAGE in amphoteric buffers

Polyacrylamide gel electrophoresis of proteins under native conditions is described at several occasions in part 1.

see pages 28 and following, as well as 35 and 57.

A series of buffers systems have been developed for the native separation of proteins (Jovin *et al.* 1970; Ornstein, 1964; Davis, 1964; Maurer, 1968). Disc electrophoresis in a basic buffer according to *Ornstein and Davis* is the technique most used. A homogeneous glycine-acetic acid buffer at pH 3.1 is used for the separation of basic proteins, but the gel preparation is relatively complicated.

Catalysts such as ammonium persulfate and TEMED are necessary for the polymerization of acrylamide. These substances dissociate in the gel and form ions which can have a marked influence on the buffer system calculated beforehand.

For these reasons many buffers could only be used in agarose gels earlier, but electroendosmosis was a problem then.

When gels are polymerized on support films and the catalysts are washed out, the gels can be dried and rehydrated in the desired buffer and the problem thus avoided .

These gels can be prepared in the laboratory or purchased in the dried form.

An amphoteric buffering substance which establishes a constant pH value in the gel near its pI can be used as for titration curve analysis (steady-state value). If the buffering substance is not influenced by other ions or electrolytes, it is not charged and cannot migrate.

see also the principles of "rocket" immunoelectrophoresis page 14.

A washed and dried homogeneous gel is simply rehydrated in the buffer solution and the electrodes are connected directly to the gel for the separation of low molecular weight non-amphoteric substances.

No buffer reservoirs are necessary.

A discontinuity in the buffer system and in the gel porosity appreciably increases the sharpness of the bands and the resolution during the separation of proteins. The amphoteric compound is only needed in the gel.

The buffer strips contain leading or terminating ions and an acid or a base to increase the conductivity and ionization of the substances.

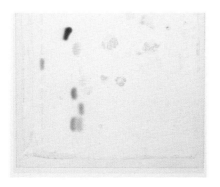

Fig. 1: Electrophoresis of cationic dyes in 0.5 mol/L HEPES, method a.

Two systems have been chosen for the following explanations:

> *a) a separation method for cationic dyes.*
> *b) a cationic electrophoresis for proteins.*

1 Sample preparation

Method a (dyes)

● Dissolve 10 mg of each dye in 5 µL of double-distilled water.   *Apply 1.5 µL*

Method b (proteins)

● Marker proteins pI 5.5 to 10.7 + 100 µL of double-distilled   *Apply 6.5 µL*
water.

● Meat extracts from pork, rabbit, veal and beef frozen in   *Apply 6.5 µL*
portions. Dilute before use: 100 µL of meat extract + 300 µL
of double-distilled water.

Other samples:

● Set the protein concentration around 1 to 3 mg/mL. Dilute   *Apply 6.5 µL*
with double-distilled water. The salt concentration should not
exceed 50 mmol/L.

It may be necessary to desalt with a NAP-10 column: apply 1
mL of sample - use 1.5 mL of eluent.

2 Stock solutions

Acrylamide, Bis solution "SEP" (T = 30%, C = 2%):
29.4 g of acrylamide + 0.6 g of bisacrylamide, make up to
100 mL with double-distilled water (H$_2$O$_{Bidist}$).

C = 2% in the resolving gel so-lution prevents the **separation** *gel from peeling off the support film and cracking during drying.*

Acrylamide, Bis solution "PLAT" (T = 30%, C= 3%):
29.1 g of acrylamide + 0.9 g of bisacrylamide, make up to
100 mL with double-distilled water (H$_2$O$_{Bidist}$).
Caution! *Acrylamide and bisacrylamide are toxic in the mono-meric form. Avoid skin contact and dispose of the remains ecologically (polymerize the remains with an excess of APS).*

This solution is used for slightly concentrated **plateaus** *with* **C = 3%***, because the slot would be-come unstable if the degree of polymerization were lower.*

Ammonium persulfate solution (APS) 40% (w/v):

Dissolve 400 mg of ammonium persulfate in 1 mL H$_2$O$_{Bidist}$.

It can be stored for one week in the refrigerator (4 °C).

0.25 mol/L Tris-HCl buffer:

3.03 g of Tris + 80 mL of double-distilled water, titrate to pH 8.4
with 4 mol/L HCl, make up to 100 mL with double-distilled water.
This buffer corresponds to the anode buffer of horizontal
electrophoresis as in method 7. It is used to set the pH of the
polymerization solution, since polymerization works best in a
slightly basic medium.

The buffer is removed during the washing step.

100 mmol/L arginine solution:

1.742 g of arginine, make up to 100 mL with double-distilled water.   *store at +4 °C*

300 mmol/L acetic acid :

1.8 mL of acetic acid (96%), make up to 100 mL with double-distilled water.

300 mmol/L ε-amino caproic acid:

18 g of ε-amino caproic acid, make up to 500 mL with double-distilled water.   *store at +4 °C*

Pyronine solution (cationic dye marker) 1% (w/v):

1 g of pyronine, make up to 100 mL with double-distilled water.

Glycerol 87%:

Glycerol fulfills many functions: it improves the fluidity of the polymerization solution and makes it denser at the same time so that the final overlayering is easier. In the last washing step, it ensures that the gel does not roll up when it is dried.

3 Preparing the empty gels

Slot former

Sample application is done in small wells which are molded in the surface of the gel during polymerization. To form these slots a mould must be fixed on a glass plate (spacer).   *The "spacer" is the glass plate with the 0.5 mm thick U-shaped silicone rubber gasket fixed to it.*

● *Method a:*

The casting mold for ultrathin gels (0.25 mm) from method 1 is used.

● *Method b:*

The cleaned and degreased glass plate with 0.5 mm U-shaped spacer is placed on the template (slot former template in the appendix). A layer of "Dymo" tape (embossing tape, 250 μm thick) is applied, avoiding air bubbles, at the starting point. Several layers of "Scotch tape" (one layer 50 μm) can be used instead (Fig. 2). The slot former is cut out with a scalpel. After pressing the individual slot former pieces against the glass plate, the remains of sticky tape are removed with methanol.   *"Dymo" tape with a smooth adhesive surface should be used. Small air bubbles can be enclosed when the adhesive surface is structured, these inhibit polymerization and holes appear around the slots.*

The casting mold is then made hydrophobic by spreading a few mL of Repel Silane over the whole slot former with a tissue under the fume hood. When the Repel Silane is dry, the chloride ions which result from the coating are rinsed off with water.   *This treatment only needs to be carried out once*

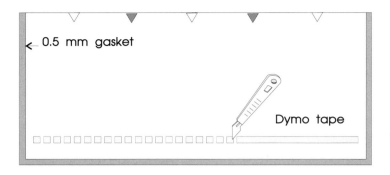

0.5 mm gasket

Dymo tape

Fig. 2: Preparing the slotformer.

Assembling the casting cassette

For mechanical stability and to facilitate handling, the gel is covalently polymerized on a support film. The glass plate is placed on an absorbent tissue and wetted with a few mL of water. The GelBond PAG film is applied with a roller with the untreated hydrophobic side down (Fig. 3). A thin layer of water then forms between the film and the glass plate and holds them together by adhesion. The excess water which runs out is soaked up by the tissue. To facilitate pouring in the gel solution, the film should overlap the length of the glass plate by about 1 mm.

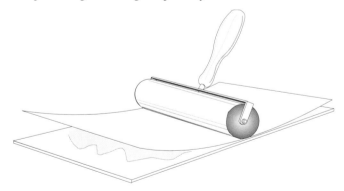

Fig. 3: Rolling on the support film.

The finished slot former is placed on the glass plate and the cassette is clamped together (Fig. 4).

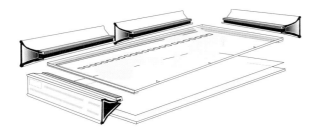

Fig. 4: Assembling the casting cassette.

Polymerization solutions

● *Method a: gel recipe for 2 gels (T = 8%, C = 3)*

Introduce and mix in test tubes with screw caps (15 mL):
4.0 mL of acrylamide, Bis solution "PLAT"
0.5 mL of Tris-HCl
make up to 15 mL with double-distilled water
7 µL of TEMED (100%)
15 µL of APS

Increasing T leads to less sharp bands.

See method 1 for the casting technique. These gels are also washed and dried.

● *Method b: discontinuous gel*

Cool the casting cassette in the refrigerator at 4 °C for about 10 min: this delays the onset of polymerization. This step is necessary because the stacking gel with large pores and the resolving gel with small pores are cast in one piece. The polymerization solutions which have different densities take 5 to 10 min to settle.

In the summer in a warm laboratory, the gel solutions should also be brought to 4 °C.

Composition of the polymerization solutions:

Stacking gel (T = 4%, C = 3%)	Resolv.gel (T = 10%, C = 2%)
1.3 mL acrylamide/Bis"PLAT"	5.0 mL acrylamide/Bis "SEP"
0.2 mL of Tris-HCL	0.3 mL of Tris-HCl
2 mL of glycerol	0.3 mL of glycerol
5 µL of TEMED	7 µL of TEMED
make up to 10 mL with	make up to 15 mL with dou-
double-distilled water	ble-distilled water
+ 10 µL of APS	+ 15 µL of APS
Introduce *3 mL*	pipette *13* mL
in the cassette	on the solution

APS is only added shortly before filling the cassette.

Filling the cooled gel cassette

The cassette is filled with a 10 mL pipette or a 20 mL syringe (Fig. 5). Draw the solution into the pipette with a pipetting device. The stacking gel plateau is introduced first, and then the resolving gel which contains less glycerol and is less dense. Pour the solutions in slowly. The gel solution is directed into the cassette by to the piece of film sticking out.

Never pipette the toxic mono-mers by mouth!

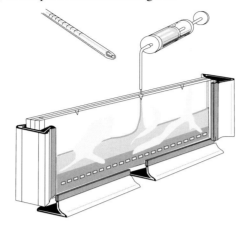

Fig. 5: Introducing the polymerization solutions in the gel cassette.

100 μL of 60 % (v/v) isopropanol are then layered in each filling notch. Isopropanol prevents oxygen, which inhibits poly-merization, from diffusing into the gel. The gel will then present a well-defined, aesthetic upper edge.

If air bubbles are trapped in the solution, they can be dislodged with a long strip of polyester film.

Let the gel stand at room temperature .

Polymerization

Removing the gel from the casting cassette

After the gel has polymerized over night, the clamps are removed and the glass plate gently lifted off the film with a spatula. The gel can slowly be pulled away from the spacer by grasping a corner of the film.

There is a slow "silent polyme-rization" after the gel has solidi-fied, which should be completed before the gel is washed.

Washing the gel

The gels (0.25 mm gels for dyes and 0.5 mm gels for proteins) are washed three times for 20 min by shaking in double-distilled water. The last washing solution should contain 2% glycerol. If a "MultiWash" apparatus is used the gel is deionized by pumping double-distilled water through the mixed bed ion-exchanger cartridge for 30 min (Fig. 6).

The remains of monomers, APS and TEMED are removed from the gel.

● Before drying the gel, soak it for about 15 min in a 10% glycerol solution.

● Air-dry the gel overnight.

Then store them in the refrigera-tor covered with polyester film.

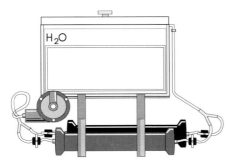

Fig. 6: MultiWash. Gel washing and destaining appliance.

4 Electrophoresis

Rehydration in amphoteric buffers

● Lay GelPool onto a horizontal table; select the appropriate reswelling chamber, pipet rehydration solution into the chamber, for
a complete gel: 25 mL
a half gel: 13 mL
● Set the edge of the gel-film – with the gel surface facing down – into the rehydration buffer (Fig. 7) and slowly lower it, avoiding air bubbles.
● Using forceps, lift the film up to its middle, and lower it again without catching air bubbles, in order to achieve an even distribution of the liquid (Fig. 7 B). Repeat this during the first 10 min.

The Gels can be used in one piece, or – depending on the number of samples – cut into smaller portions with scissors (when they are still dry). The rest of the gel should be sealed airtight in a plastic bag and stored in a freezer.

Very even rehydration is obtained when performing it on a shaker at a slow rotation rate (Fig. 7C). If no shaker is used, lift gel edges repeatedly.

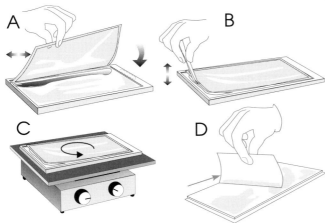

Fig. 7: Rehydration of a gel.
A Placing the dry gel into the GelPool. **B** Lifting the gel for an even distribution of the liquid. **C** Rehydration on a rocking platform (not always necessary). **D** Removing the excess buffer from the gel surface with filter paper.

60 min later the gel has reswollen completely and is removed from the GelPool. Dry sample wells with clean filter paper, wipe buffer off the gel surface with the edge of a filter paper (Fig. 7D).

When the gel surface is dry enough, this is indicated by a noise like a whistle.

Method a (dyes):

Rehydration solution (0.5 mol/L HEPES)

● 1.15 g of HEPES, make up to 10 mL with H_2O_{Bidist}.

The very thin gel needs only 10 mL of reswelling solution

● Rehydrate for about 1 h.
● Dry the sample wells with filter paper.
● Wet the cooling plate with 1 mL of *contact fluid* either kerosene or DC-200 silicone oil.

Water or other liquids are not suitable, as they can cause electrical shorting.

● Place the gel, film side down, on the cooling plate.

Avoid air bubbles.

In this method the electrodes can be placed directly on the edge of the gel; no buffer wicks are necessary.

see methods 3 and 6

● Quickly introduce the samples into the wells:

1.5 μL in each,

● Power supply: 400 V_{max}, 60 mA_{max}, 20 W_{max}, about 1 h.

After separation

● Switch off the power supply and open the safety lid.

Immediately place the dye gel on a warm surface, a light box for example. A separation is shown on Fig. 1.

It is important to dry the gel immediately for mechanical fixing of the zones.

Method b (PAGE of cationic proteins):

For the separation of lipophilic proteins (e.g. alcohol soluble fractions from cereals), it suggested to add 0.1 to 0.5 % (w/v) ProSolv II and 1 to 3 mol/L urea to the rehydration buffer.

add urea and detergents only for these applications. with lipophilic membrane proteins.

Rehydration solution (0.6 mol/L HEPES):

0.6 mol/L HEPES	3.5 g
1 mmol/L acetic acid	83 μL (from the stock solution)
10 mmol/L arginine	2.5 mL (from the stock solution)
0.001 % pyronine	8μL (from the stock solution)

The dye pyronine permits the front to be seen.

make up to 25 mL with double-distilled water.

Rehydration time: about 1 h.

● Dry out the sample wells and the gel surface with the edge of a clean filter paper.

Cathode buffer:	*Anode buffer:*
30 mmol/L acetic acid:	113 mmol/L ε-aminocaproic
2 mL (from the stock solution)	acid: 7.5 mL
	(from the stock solution)
	5 mmol/L acetic acid: 333 μL
	(from the stock solution)
to → 20mL with H_2O_{bidist}	*to → 20mL with H_2O_{bidist}*

● Wet the cooling plate with 1 mL of *contact fluid,* either kerosene or DC-200 silicone oil.

Water or other liquids are not suitable, as they can cause electrical shorting.

● Place the gel (surface up) onto the center of the cooling plate: the side containing the sample wells must be oriented towards the anode (Fig. 9).

Multiphor II: cathodal side of the wells matching line no. 12

● Lay two of the electrode wicks into the compartements of the PaperPool (if smaller gel portions are used, cut them to size). Apply 20 mL of the anode and cathode buffer respectively to the wicks (Fig. 8). Place the anode strip onto the anodal edge of the gel, matching the grid on the cooling plate between the lines 13 and 14. Place the cathode strip onto the cathodal edge, matching the grid between 3 and 4 (Fig. 9).

Always apply anode wick first, in order to prevent buffer contamination of the trailing ions.

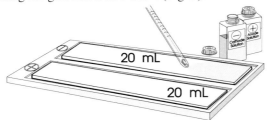

Fig. 8: Soaking the wicks with electrode buffer.

● Smooth out air bubbles by sliding bent tip foreceps along the edges of the wicks laying in contact with the gel (first anode, then cathode).

● Quickly fill the sample wells:

6.5 µL in each.

● Move electrodes so that they will rest on the outer edge of the electrode wicks. Connect the cables of the electrodes to the apparatus and lower the electrode holder plate (Fig. 9). Close the safety lid.

Clean platinum electrode wires before (and after) each electrophoresis with a wet tissue paper.

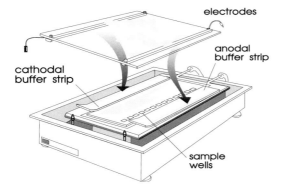

Fig. 9: Cationic native electrophoresis: sample application at the anode.

Running conditions: (10 °C) (maximum settings):

Tab. 1: Power supply program.

	U (V)	I (mA)	P (W)	t (min)
complete gel	500 V	10 mA	10 W	10 min
	1200 V	28 mA	28 W	50 min
half gel	500 V	5 mA	5 W	10 min
	1200 V	14 mA	14 W	50 min

Phase 1 is for a mild sample entry and effective stacking.

The field strength can be increased by low conductivity: ™ fast separation for a native system.

If composite proteins such as chloroplast proteins are separated, the voltage should not exceed 300 V.

The separation will then take more time.

After separation

● Switch off the power supply and open the safety lid.

● Remove the electrode wicks and either place them on the edge of the cooling plate or let them slip into the tank.

● Protein gels are stained or blotted.

see method 9 for blotting.

5 Coomassie and silver staining

Colloidal Coomassie staining

The results are quickly visible with this method. Few steps are needed, the staining solutions are stable, there is no background staining. Oligopeptides (10 to 15 amino acids) which are not properly fixed by other methods are detected. In addition, the solution is almost odorless (Diezel *et al* 1972; Blakesley and Boezi, 1977).

The run is complete, when the buffer front (pyronin dye marker) has reached the anodal strip.

Preparation of the staining solution:
Dissolve 2 g of Coomassie G-250 in 1 L of distilled water and add 1 L of sulfuric acid (1 mol/L or 55.5 mL of concd H_2SO_4 per L) while stirring. Filter (folded filter) after stirring for 3 h, add 220 mL of sodium hydroxide (10 mol/L or 88 g in 220 mL) to the brown solution. Finally add 310 mL of 100% TCA (w/v) and mix well. The solution will turn green.

Fixing and staining: 3 hours at 50 °C or at room temperature overnight in the colloidal solution;
Washing out the acid: soak in water for 1 or 2 hours, the green bands will become blue and more intense.

Fast Coomassie staining

Staining: hot staining by stirring with a magnetic stirrer in the stainless steel staining tank with 0.02 % Coomassie R-350: dissolve 1 PhastGel Blue tablet in 2 L of 10% acetic acid; 8 min, 50 °C (see Fig. 10);

The gel is placed on the grid with the upper surface down.

Destaining: in the MultiWash in 10% acetic acid for 2 h at room temperature (see Fig. 11);

The MultiWash is composed of a destaining tank, a pump, and charcoal, mixed bed ion-exchange cartridges.

Preserving: in a solution of 25 mL of glycerol (87% w/v) + 225 mL of distilled water for 30 min;

Drying: air-dry (room temperature).

5 minute silver staining of dried gels

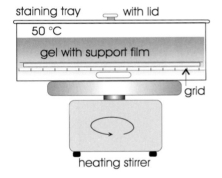

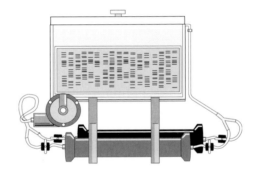

Fig. 10: Appliance for hot staining of gels.   **Fig. 11:** Destaining in the MultiWash

This method is used for staining dried gels already stained with Coomassie (the background must be completely clear) so as to increase the sensitivity. It can also be used to for direct staining of gels after pretreatment (Krause and Elbertzhagen, 1987). A particular advantage of this method is that no proteins and peptides are lost. They frequently diffuse out of the gel during other silver staining methods because they are not irreversibly fixed in focusing and native gels as during SDS electrophoresis.

Pretreatment of unstained gels:

● fix for 30 min in 20% TCA,

● wash 2 x 5 min in 45% methanol, 10% glacial acetic acid,

● wash 4 x 5 min in distilled water,

● impregnate for 2 min in 0.75% glycerol,

● air-dry.

Solution A: 25 g of Na_2CO_3, 500 mL of double-distilled water;
Solution B: 1.0 g of NH_4NO_3, 1.0 g of $AgNO_3$, 5.0 g of tungsto-silisic acid, 7.0 mL of formaldehyde solution (37%), make up to 500 mL with double-distilled water.

Silver-staining:

Before use mix 35 mL of solution A with 65 mL of solution B. Immediately soak the gel in the resulting whitish solution and incubate while agitating until the desired intensity is reached. Rinse briefly with distilled water.

Stop with 0.05 mol/L glycerol. Remove the metallic silver from the gel and the support with a cotton swab.

Air-dry.

Fig. 12 shows an electropherogram of different protein samples which were analyzed as described here.

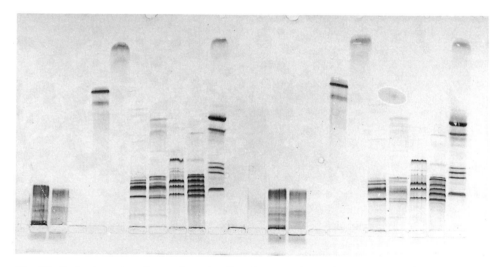

Fig. 12: Cationic native electrophoresis of marker proteins and meat extracts in 0.6 mol/L HEPES. Staining with PhastGel Blue R-350 (Cathode on top.).

For *alkaline* separations, no amphoteric buffer has been found yet, which meets the buffer capacity requirements. Very good results are obtained with washed and dried gels rehdrated in the conventional Tris-HCl buffer for basic native electrophoresis.

In the cathode Tris-glycine is used in this case.

Method 5: Agarose IEF

The principles of isoelectric focusing in carrier ampholyte pH gradients are described in part I. There are several reasons for the use of agarose as separation medium.

See pages 45 and following.

● No toxic monomer solutions are needed;

No acrylamide, no Bis.

● there are no catalysts which can interfere with the separation;

No TEMED, no APS.

● no polymerization solutions have to be prepared and stored;

Less work.

● the matrix has large pores;

High molecular weight proteins are not a problem.

● therefore the separation times are shorter;

Less frictional resistance.

● the staining times are shorter; immunofixation can be carried out in the gel.

The dried gel is stained.

See page 12.

There are however, a few problems with agarose:

● the gels are not absolutely free of electroendosmosis, since not all the carboxyl and sulfate groups are removed during washing.

This leads to a cathodic drift of the gradient and transport of water in the gel.

● for these reasons agarose gels are not suited for use at extreme pH ranges (acid or basic).

Agarose electrophoresis works best in the middle of the pH range.

● gels containing urea are difficult to prepare, because urea disrupts the agarose network.

Use orehydratable agarose gels (Hoffman et al. 1989).

1 Sample preparation

● Marker proteins pI 4.7 to 10.6 or

● Marker proteins pI 5.5 to 10.7 + 100 µL of double-distilled water.

Apply 10 µL.

● Meat extracts from pork, rabbit, veal and beef frozen in portions. Dilute before use: 100 µL of meat extract + 300 µL of double-distilled water.

Apply 10 µL.

● Other samples:

Set the protein concentrations around 1 to 3 mg/mL. Dilute with double-distilled water. The salt concentration should not exceed 50 mmol/L.

Apply 10 µL.

Desalting with a NAP-10 column may be necessary: apply 1 mL – use 1.5 mL of eluent.

2 Preparing the agarose gel

Agarose gels are cast on GelBond film: a polyester film coated with a dry agarose layer.

not to be confused with Gel-Bond PAG film.

Agarose gels can be cast several ways: oxygen from the air does not inhibit gelation.

unlike polymerization of acrylamide.

Here the gel is cast in a vertical prewarmed cassette, since this produces an uniform gel layer. The following method is based on the experience of Dr. Hans-Jürgen Leifheit, Munich (1987) whom we would like to thank for his helpful advice.

Leifheit H-J, Gathof AG, Cleve H. Ärztl Lab. 33 (1987) 10-12.

Making the spacer hydrophobic

A few mL of Repel Silane are spread over the inner face of the spacer with a tissue under the fume hood. When the Repel Silane is dry, the chloride ions which result from the coating are rinsed off with water.

The "spacer" is the glass plate with the 0.5 mm thick U-shaped silicone rubber gasket.
This treatment only needs to be carried out once.

Assembling the casting cassette:

● Remove the GelBond film from the package.

Identify the hydrophilic side with a few drops of water.
This facilitates filling the cassette later on.

Pour a few mL of water on the glass plate and place the support film on it with the hydrophobic side down. Press the film against the glass with a roller (Fig. 1), the film should overlap the length of the glass plate by about 1 mm.

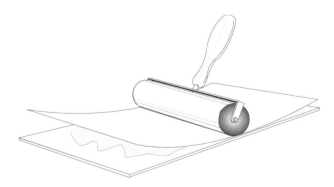

Fig. 1: Applying the support film with a roller.

The spacer, with the gasket on the bottom, is then placed on the glass plate and the cassette is clamped together (Fig. 2).

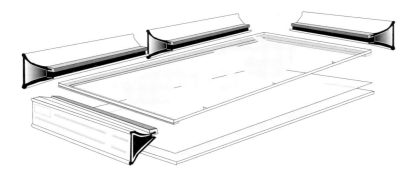

Fig. 2: Assembling the gel cassette.

Before pouring the hot agarose solution, prewarm the cassette and a 10 mL pipette to 75 °C in a heating cabinet.

This ensures that the solution does not solidify immediately.

Preparation of the agarose solution (0.8% agarose)

It is important to store agarose in a dry place because it is very hygroscopic.

If the agarose is humid, too little will be weighed.

● In a 100 mL beaker:
2 g of sorbitol
19 mL of double-distilled water
0.16 g of agarose IEF

Sorbitol improves the mechanical properties of the gel and since it is hygroscopic, it works against the electro-osmotic water flow.

● Mix and boil the solution – covered with a watch glass – until the agarose has completely dissolved either in a microwave oven at the lowest setting or while stirring slower on a magnetic heater.

Rapid stirring damages the mechanical properties of agarose, which is why heating in the microwave oven is preferable.

● Degas to remove CO_2.

● Place the beaker in a heating cabinet for a few minutes to cool the solution down to 75 °C.

It should not be too hot for the carrier ampholytes.

● Add 1.3 mL of Ampholine pH 3.5 to 9.5 or Pharmalyte pH 3 to 10 and stir with a glass rod.

● Remove the cassette from the heating cabinet; draw the hot agarose solution in the prewarmed pipette and quickly release it in the cassette (Fig. 3).

Avoid air bubbles. Should some be trapped nevertheless, dislodge them with a long strip of polyester film.

● Let the cassette stand for 1 or 2 h at room temperature.

The gel slowly sets.

● Remove the clamps and take out the gel.

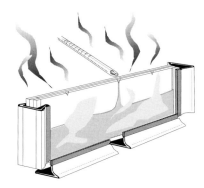

Fig. 3: Pouring the hot agarose solution in the prewarmed cassette.

● Place the gel on a piece of moist tissue and store it overnight in a humidity chamber (Fig. 4) in the refrigerator, it can be kept for up to a week.

Only then does the final agarose gel structure form (see page 8).

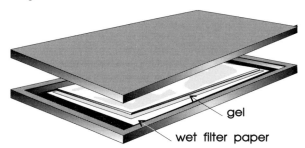

gel

wet filter paper

Fig. 4: Agarose gel in the humidity chamber.

Preparation of electrode solutions

In table 1 a list of electrode solutions is given for respective pH intervals.

Tab. 1: Electrode solutions for IEF in agarose gels.

pH Gradient	Anode	Cathode
3.5 – 9.5	0.25 mol/L acetic acid	0.25 mol/L NaOH
2.5 – 4.5	0.25 mol/L acetic acid	0.40 mol/L HEPES
4.0 – 6.5	0.25 mol/L acetic acid	0.25 mol/L NaOH
5.0 – 8.0	0.04 mol/L glutamic acid	0.25 mol/L NaOH

3 Isoelectric focusing

● The gel is placed with the film on the bottom on the cooling plate at 10 °C, using about 1 mL kerosene or DC-200 silicone oil (Fig. 6).

IEF must be performed at a defined constant temperature because the pH gradient and the pIs are temperature dependent.

● Dry the surface with filter paper (Fig. 5) because agarose gels have a liquid film on the surface.

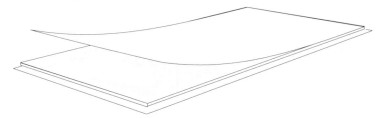

Fig. 5: Drying the surface of the gel with filter paper.

Electrode solutions:

To reduce cathodic drift apply electrode strips made from filter paper between the electrodes and both edges of the gel and let them soak in the electrode solutions. For a pH gradient from 3.5 to 9.5 these are:

Anode:	Cathode:
0.25 mol/L acetic acid	0.25 mol/L NaOH

These are also used for pH 4.0 to 6.5.

Cut the electrode strips shorter than the gel (< 25 cm);

They should not hang over the sides.

● soak the strips thoroughly in the corresponding solutions;

● blot them with dry filter paper for about 1 min to remove excess liquid;

● place the electrode strips along the edge of the gel;

● shift the electrodes along the electrode holder so that they lie over the electrode strips (see Fig. 6);

Make sure that the acid strips lies under the anode and the basic strip under the cathode.

● plug in the cable, make sure that the long anodic connecting cable is hooked to the front.

Separation conditions
The values for current and power are valid for a whole gel. For IEF of half a gel use half the values for mA and W.

● *Prefocusing:* 30 min at max. 1400 V, 30 mA, 8 W,

The pH gradient is formed during this step.

● *Sample application:* apply 10 mL in each hole 2 cm away from the anode.

*Do **not** use any filter paper or cellulose plates for agarose IEF!*

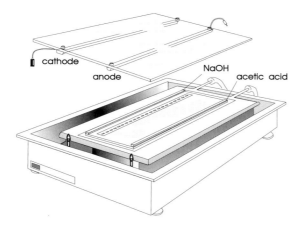

cathode

anode

NaOH

acetic acid

Fig. 6: Agarose IEF with electrode strips and sample application strip.

The optimum point for sample application in agarose IEF depends on the characteristic of the sample (see method 6). But most samples can be applied where mentioned here.

It may be necessary to do a step test.

● Desalting: 30 min at max. 150 V; the other settings remain.

This also helps high molecular weight proteins, e.g. IgM to enter the gel.

● Separation: 60 min at max. 1500 V, 30 mA, 8 W.

This is valid for gradients from pH 3 to 10.

During the run it may be necessary to interrupt the separation and blot the electrode strips with filter paper:

For narrower gradients, e.g. pH 5 to 8, IEF should last about 2 h, since the proteins with a low net charge must cover long distances.

● switch off the power supply, open the safety lid, remove the electrodes and blot the excess water from the strips;

● continue the separation.

The proteins are then stained, immunofixed or blotted. Should problems occur, consult the trouble-shooting guide in the appendix.

4 Protein detection

Coomassie Blue staining

Prepare all solutions with distilled water.

● *Fixing:* 30 min in 20% (w/v) TCA;
● *Washing:* 2 × 15 min in 200 mL of fresh solution each time: 10% glacial acetic acid, 25% methanol;
● *Drying:* place 3 layers of filter paper on the gel and a 1 to 2 kg weight on top (Fig. 7). Remove everything after 10 min and finish drying in the heating chamber;

First moisten the piece of filter paper lying directly on the agarose.

glass plate

gel   filter paper

Fig. 7: Pressing the agarose gel.

● *Staining:* 10 min in 0.5% (w/v) Coomassie R-350 in 10% glacial acetic acid, 25% methanol: dissolve 3 PhastGel Blue R tablets (0.4 g dye each) in 250 mL;
● *Destaining:* in 10% glacial acetic acid, 25% methanol till the background is clear;
● *Drying:* in the heating cabinet.

Immunofixation

Only the protein bands which have formed an insoluble immunoprecipitate with the antibody are stained with this method.

All the other proteins and excess antibodies are washed out of the gel with a NaCl solution.

Antibody solution: dilute 1:2 (or 1:3 depending on the antibody titer) with double-distilled water and spread it over the surface of the gel with a glass rod or a pipette: use about 400 to 600 µL;

● *Incubation:* 90 min in the humidity chamber in a heating chamber or incubator at 37 °C;
● *Pressing:* 20 min under 3 layers of filter paper, a glass plate and a weight (Fig. 7);

First moisten the piece of filter paper lying directly on the gel.

● *Washing:* in a physiological salt solution (0.9% NaCl w/v) overnight;
● *Drying:* see Coomassie staining;
● *Staining* and destaining as for Coomassie staining.

It is important to know the antibody titer, since hollow bands can appear in the middle when the antibody solutions are too concentrated: one antigen binds to one antibody and no precipitate is formed.

Silver staining

If the sensitivity of Coomassie staining is not sufficient, silver staining is carried out on the dry gel (Willoughby and Lambert, 1983):

Solution A: 25 g of Na_2CO_3, 500 mL of double-distilled water;

Solution B: 1.0 g of NH_4NO_3, 1.0 g of $AgNO_3$, 5.0 g of tungstosilicic acid, 7.0 mL of formaldehyde solution (37%), make up to 500 mL with double-distilled water.

Staining: mix 35 mL of solution A with 65 mL of solution B just before use. Soak the gel immediately in the resulting whitish solution and incubate while agitating until the desired intensity is reached. Rinse briefly with distilled water;

Stop with 0.05 mol/L glycerol. Remove remains of metallic silver from the gel and the back of the support film with a cotton swab.

Drying: air-dry.

Method 6: PAGIEF in rehydrated gels

The principles of isoelectric focusing in carrier ampholyte pH gradients are described in part I.

see pages 45 and following.

The use of washed, dried and rehydrated polyacrylamide gels is advantageous for a number of reasons:

- A few carrier ampholytes inhibit the polymerization of gels, especially the reaction on the film; this means that the thin gel may swim around like a jellyfish in an aggressive staining solution.

  especially strongly basic ampho-lytes.

- Gels containing ampholytes are slightly sticky, and thus not very easy to remove from the casting cassette.

  because of inhibited polymeriza-tion as well.

- Blank gels can be washed: APS, TEMED and unreacted acrylamide and Bis monomers can thus be washed out of the gel. This allows focusing to occur with fewer interferences, as is readily apparent in the straight bands in the acid pH range as well.

  During IEF the otherwise indis-pensable electrode wicks, soa-ked in acid or basic buffer, can be left out: this allows separati-on up to the platinum electrode!

- Gels can easily be prepared in large quantities and kept dry.

  with fresh acrylamide monomer solution.

- Chemical additives, which allow the separation of many proteins but which would inhibit polymerization, can be added to the gel without any problems.

  for example non-ionic deter-gents such as Triton or NP-40.

- Sample wells can be formed in the gel without disturbing the pH gradient.

  see method 1 for preparation of a "slot former" for example.

- Blank, ready-made gels can be bought.

  Handling acrylamide monomers is not necessary and a lot of time and effort are saved.

1 Sample preparation

- Marker proteins pI 4.7 to 10.6 or
- Marker proteins pI 5.5 to 10.7 + 100 µL of double-distilled water.

  Apply 10 µL.

- Meat extracts form pork, rabbit, veal and beef frozen in portions. Dilute before use: 100 µL of meat extract + 300 µL of double-distilled water.

  Apply 10 µL.

- *Other samples:* Set the protein concentration around 1 to 3 mg/mL. Dilute with double-distilled water. The salt concen-tration should not exceed 50 mmol/L.

  Apply 10 to 20 µL.

It may be necessary to desalt with a NAP-10 column: apply 1 mL sample solution – use 1.5 mL of eluent.

2 Stock solutions

Acrylamide, Bis (T = 30%, C = 3%):

29.1 g of acrylamide + 0.9 g of Bis, make up to 100 mL with double-distilled water or

PrePAG Mix (29.1 : 0.9), reconstitute with 100 mL of double-distilled water.

Dispose of the remains ecologically: polymerize with an excess of APS.

Caution! *Acrylamide and Bis are toxic in the monomeric form. Avoid skin contact and do not pipette by mouth.*

When stored in a dark place at 4 °C (refrigerator) the solution can be kept for one week.

It can still be kept for several weeks for SDS gels.

Ammonium persulfate solution (APS) 40% (w/v):

dissolve 400 mg of ammonium persulfate in 1 mL of double-distilled water.

stable for one week when stored in the refrigerator (4 °C).

0.25 mol/L Tris-HCl buffer:

3.03 g of Tris + 80 mL of double-distilled water, titrate to pH 8.4 with 4 mol/LHCl, make up to 100 mL with double-distilled water.

This buffer corresponds to the anodic buffer for horizontal electrophoresis as in method 7. It is used to set the pH of the polymerization solution, since it works best at a slightly basic pH.

The buffer is removed during the washing step.

Glycerol 87% (w/v):

Glycerol fulfills several functions: it improves the fluidity of the polymerization solution and makes it denser at the same time so that the final overlayering is easier. In the last washing solution, it ensures that the gel does not roll up during drying.

3 Preparing the blank gels

Assembling the casting cassette

● Remove the GelBond PAG film from the package.

Identify the hydrophilic side with a few drops with water.

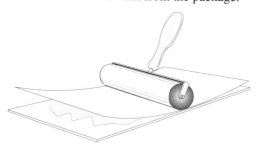

Fig. 1: Applying the support film with a roller.

Pour a few mL of water on the glass plate and place the support film on it with the hydrophobic side on the bottom. Press the film on the glass plate with a roller so that the film overlaps the length of the glass plate by about 1 mm.

This facilitates filling the cassette later on.

Since the gel should bind to the GelBond PAG film after polymerization but not to the spacer, the spacer is made hydrophobic with Repel Silane. It is then placed on the film with the gasket on the bottom and the cassette is clamped together (Fig. 2).

The "spacer" is the glass plate with the 0.5 mm thick U-shaped silicone rubber gasket.

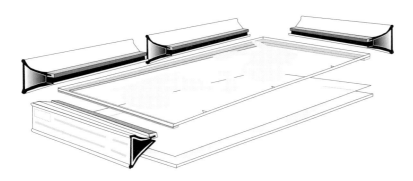

Fig. 2: Assembling the casting cassette.

● Composition of the polymerization solution for a gel with $T = 4.2\%$:

2.7 mL of acrylamide/Bis
0.5 mL of Tris-HCl
1 mL of glycerol
10 µL of TEMED
make up to 20 mL with double-distilled water.

Filling the casting cassette

When everything else is ready, add 20 mL of APS to the polymerization solution.

The clock is running now: this solution polymerizes after at most 20 min.

The cassette is filled with a 10 mL pipette or a 20 mL syringe (Fig. 4). Fill the pipette completely (the 10 mL scale plus the headspace make 18 mL) and place the tip of the pipette in the middle notch of the spacer. Release the solution slowly. The gel is directed in to the cassette by the piece of film sticking out.

Fig. 3: Pouring the polymerization solution.

100 µL of 60 % v/v isopropanol-water are then pipetted into the 3 notches. Isopropanol prevents oxygen which inhibits polymerization from diffusing in the gel. The gel then presents a well defined aesthetic edge.

Air bubbles which might be trapped can be dislodged with a long strip of polyester film.

Let the gel stand for one hour at room temperature.

Polymerization.

Removing the gel from the casting cassette

After the gel has polymerized for one hour, remove the clamps and carefully lift the glass plate from the support film with a spatula. Slowly peel the gel away from the spacer by pulling on a corner of the film.

Washing the gel

Wash the gel three times with shaking in double-distilled water for 20 min. The last washing step should contain 2% glycerol. If a MultiWash apparatus is used: deionize by pumping double-distilled water through the mixed bed ion-exchange cartridge for about 30 min (Fig. 4).

This step washes the remains of monomers, APS and TEMED out of the gel.

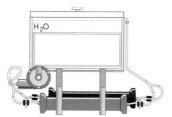

Fig. 4: MultiWash. Gel washing and destaining appliance.

Drying the gel

Dry the gels overnight at room temperature with a fan. Then cover them with a protecting film, and store them frozen (−20 °C) in sealed plastic bags.

Heat drying damages the swelling capacity, which is why the gels should be stored frozen.

4 Isoelectric focusing

Rehydration solution (Ampholine, Pharmalyte):

7.9 mL of ethylenglycol (= 7.5 % v/v)
200 µL of Ampholine + 580 µL Pharmalyte
make up to 10.5 mL with double-distilled water.
Narrow gradients: When narrow gradients (e.g. pH 5 to 7) are
employed, add 10 % (v/v) carrier ampholytes of a wide range
(e.g. 3 to 10). See also "7 Perspectives" on page 163 ff.

Example for an optimized broad gradient pH 3 to 10.
7.5 % (v/v) of ethylenglycol in the gel increases viscosity, so the bands are straighter.

Focus narrow gradients for a longer time: separation phase prolongued up to 3 hours.

Reswelling the gel

The gels can be used in one piece or cut into two halves or
smaller portions – depending on the number of samples to be
separated.

We suggest to degas the solution in order to remove CO2 for sharper bands in the alkaline region.

Place the GelPool on a horizontal table. Select the appropriate
reswelling chamber of the GelPool, depending on the gel size.
Clean it with distilled water and tissue paper. Pipet the appropriate volume of rehydration solution, for

complete gel:	10.4 mL
half gel:	5.2 mL

Set the edge of the gel-film – with the dry gel surface facing
downward – into the rehydration solution (Fig. 5A) and
slowly lower the film down.

At the same time move the gel-film to and fro, in order to achieve an even distribution of the liqid and to avoid trapping airbubbles.

Lift the film at the edges with forceps, and slowly lower them
down, in order to maintain an even distribution of the liquid
(Fig. 5B) and to remove air bubbles.

Note: *Repeat this measure several times during the first 15 min, to prevent the gel from stikking to the GelPool. Check, whether the gel can be moved around on its reswelling liquid.*

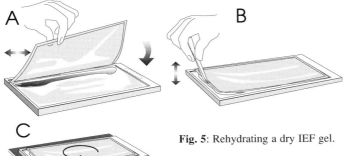

Fig. 5: Rehydrating a dry IEF gel.

A Laying the gel surface into the tray of the GelPool, which contains the exact amount of rehydration solution, which is needed.
B Lifting the edges of the gel with forceps to prevent sticking of
the gel to the tray surface.
C Rehydrating on a rocking platform (not absolutely necessary).

60 min later the gel has taken up the complete volume of the
solution and can be removed from the GelPool.

If the solution contains urea or nonionic detergents, reswelling takes longer time.

Separation of proteins

● Cool the cooling plate to 10 °C and place the gel on it, with the film on the bottom, using a few mL of kerosene or DC-200 silicone oil (Fig. 6).

IEF must be performed at a constant defined temperature since the pH gradient and the pIs are temperature dependent.

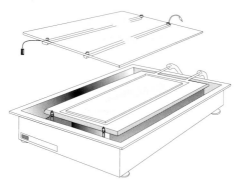

Fig. 6: The focusing electrodes are directly placed on the edges of the gel.

● The following settings for separation have been optimized:

Power Supply   Maximum settings   Time of IEF

Complete Gel:

	700 V	12 mA	8 W	20 min	(pre-IEF)
	500 V	8 mA	8 W	20 min	(sample entrance)
(half gel:	2000 V	14 mA	14 W	90 min	(separation)
1/2 mA and 1/2 W)	2500 V	14 mA	18 W	10 min	(band sharpening)

● Place the electrodes directly on the edges of the gel (Fig. 6).

Electrode wicks do not have to be used for washed gels.

● Plug in the cable; make sure that the long anodic connecting cable is hooked to the front.

● First focus without sample. This already distributes the ampholytes in the pH gradient without the proteins migrating. In addition there are proteins which are only stable in an acidic or basic medium and before prefocusing the gel has a pH value of about 7.0 all over.

Prefocusing is not necessary for all samples, since APS and TE-MED are already washed out and most proteins are not sensitive to pH; it is best to try out.

Sample application

During application it should be taken into consideration that most proteins have an optimum application point. If necessary it should be determined by applying the sample at several places (step test, Fig. 7). The samples are applied with small pieces of filter paper or silicone rubber applicator mask.

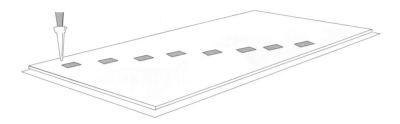

Fig. 7: Step trial test to determine the best application point.

The time for separation is sufficient for most of the proteins. Large proteins may require slightly longer to reach their pI. Viscous gels (with 20% glycerol or 8 mol/L urea solution) also require a longer focusing time.

This is valid for pH gradients from 3 to 10. When gels with a narrower pH range are used, e.g. pH 5 to 7, focusing should be performed during approximately 4 h since the proteins with lower net charge must cover longer distances.

● After IEF, switch off the power supply and open the safety lid.

● The proteins are now stained or blotted. Should problems arise, consult the trouble-shooting guide in the appendix.

5 Coomassie and silver staining

Colloidal Coomassie staining

The results are quickly visible with this method. Few steps are required, the staining solutions are stable and there is no background staining. Oligopeptides (10 to 15 amino acids) which are not sufficiently fixed by other methods can be identified here. In addition, the solution is almost odorless (Diezel *et al* 1972; Blakesley and Boezi, 1977).

Preparation of the staining solution:

Dissolve 2 g of Coomassie G-250 in 1 L of distilled water and add 1 L of sulfuric acid (1 mol/L or 55.5 mL of conc. H_2SO_4 per L) while stirring. After stirring for 3 h, filter (filter paper), add 220 mL of sodium hydroxide (10 mol/L or 88 g in 220 mL) to the brown filtrate. Finally add 310 mL of 100% (w/v) TCA and mix well. The solution will turn green.

Fixing, staining: 3 h at 50 °C or overnight at room temperature in the colloidal solution.

Washing out the acid: soak in water for 1 to 2 h, the green bands will become blue and more intense.

Quick Coomassie staining

Stock solutions: *all solutions with distilled water*

TCA:	100% TCA (w/v) 1 L
A:	0.2% (w/v) CuSO$_4$ + 20% glacial acetic acid
B:	60% methanol
C:	dissolve 1 tablet of Phast Blue R in 400 mL of
	H$_2$O$_{dist}$, add 600 mL methanol, stir 5 to 10 min.

1 tablet = 0.4 g of Coomassie Brilliant Blue R-350

Staining:

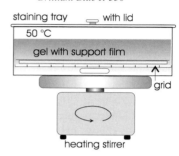

- *Fixing:* 10 min in 300 mL of 20% TCA;

- *Washing:* 2 min in 300 mL (mix A and B 1:1);

- *Staining:* 15 min in 300 mL staining solution
  (mix A and C 1:1) at 50 °C while stirring (Fig 8);

- *Destaining:* 15 to 20 min (mix A and B 1:1)

- *Impregnating:* 10 min in 200 mL 5% glycerol,
  10% acetic acid;

- *Drying:* air-dry.

Fig. 8: Appliance for hot staining.

5 minute silver staining of dried gels

The method can be used to stain gels already stained with Coomassie (the background should be perfectly clear) to increase the sensitivity or else unstained gels can be stained directly after pretreatment (Krause and Elbertzhagen, 1987). A particular advantage of this method is that no proteins or peptides are lost. They often diffuse out of the gel during other staining methods.

However it should be noted, that the sensitivity of this method is not as high as the multi-step procedure described after this one.

Pretreatment of unstained gels:

- fix for 30 min in 20% TCA,

- wash for 2 × 5 min in 45% methanol, 10% acetic acid,

- wash for 4 × 2 min in distilled water,

- impregnate for 2 min in 0.75% glycerol,

- air-dry.

Solution A: 25 g of Na$_2$CO$_3$, 500 mL of double-distilled water;
Solution B: 1.0 g of NH$_4$NO$_3$, 1.0 g of AgNO$_3$, 5.0 g of tungsto-silisic acid, 7.0 mL of formaldehyde solution (37%), make up to 500 mL with double-distilled water.
Silver staining:

Mix 35 mL of solution A with 65 mL of solution B just before use. Immediately soak the gel in the resulting white solution and incubate while agitating until the desired intensity is reached. Briefly rinse with distilled water.

Stop with 0.05 mol/L glycine. Remove remains of metallic silver from the gel and the support film with a cotton swab.

Sensitive silver staining

Because complete removal of the carrier ampholytes from the gel with simultaneous fixing of the proteins is not so easy, isoelectric focusing gels show very often a high background after highly sensitive silver staining. However, when some fixing and washing steps are performed at elevated temperatures, dark bands are developed against a light background. The method according to Heukeshoven and Dernick (1986) has been modified for manual staining. The procedure is performed with two trays: a stainless steel tray is employed for the first steps with the gel side facing down like in figure 8 on the previous page; before the silvering step the gel is transferred to a glass tray.

Heukeshoven J, Dernick R. In: Radola BJ, Hrsg. Elektrophorese-Forum'86. (1986).

In the PhastSystem the staining chamber can be heated. When another automated gel staining apparatus is used, preheated solutions are pumped in from thermos flasks.

Tab. 1: Silver staining of isoelectric focusing gels.

acc. to Heukeshoven and Dernick (1986) modified

Step	Solution	temperature	V mL	t min
Fixing 1	20 % trichloroacetic acid (w/v)	20 °C	300	15
Fixing 2	20 % ethanol / 8 % acetic acid hot!	50 °C	600	10
Fixing 3	like -2-, but + 0.4 % glutardialdehyde	50 °C	600	10
Incubation	0.1 % sodium thiosulphate; 0.4 mol/L sodium acetate / acetic acid pH 6.5; + 0.4 % glutardialdehyde	Stirring! 50 °C	600	15
Rinsing	20 % ethanol / 8 % acetic acid	50 °C	600	10
Washing	H_2O_{dist} lay gel into a glass tray, with gel surface side up	50 °C	3×600	3×10
Silvering	0.1% $AgNO_3$ 25 µL formaldehyde (37% w/v) per 100 mL toss and watch	20 °C	100	20
Developing	2.5 % Na_2CO_3 + 65 µL formaldehyde (37% w/v) per 200 mL	20 °C 20 °C	100 100	1 8
Stop / Desilver	2 % glycine + 0.5 % EDTA disodium solution	20 °C	250	10
Impregnation	2 % glycerol solution	20 °C	200	10
Drying	air dry	room temperature		

6 Densitometric evaluation

Refer to section I* and method 7** for indications on quantitative evaluation. For IEF all the natural restrictions which must
be taken into account for SDS gels are applicable. If an external
standard is used for calibration, it is recommended to separate a
mixture of pI standards, containing known concentrations, in the
same gel.

*see pages 77 and following
**see pages 183 and following

The classification of the isoelectric points of an unknown
sample with the help of an IEF marker will be described here.
The evaluation can be performed with a laser densitometer or a
high resolution desk top scanner.

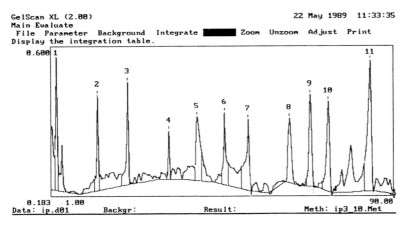

Fig. 9: Densitogram of the IEF marker pI range 3 to 10.

The gel is first placed on the light plate of the densitometer so
that the anode (of which an imprint can still be seen) lies at the
position Y = 0 of the densitometer, or is lied on the glass polate
of the desk top scanner.

After scanning (Fig. 9), the 11 pairs of coordinates – the pI of
the proteins and the electrophoretic separation distance – can be
introduced into the integration program (Fig. 10).

Measuring the unknown tracks

The unknown separation trace can now be processed. Of
course only the isoelectric points of samples which were separated in the same gel as the standard can be determined. Once the
data has been scanned (Fig. 11), the integration method   is
loaded.

Once the calculations are finished the last column contains the
classification of the isoelectric points (Fig. 12).

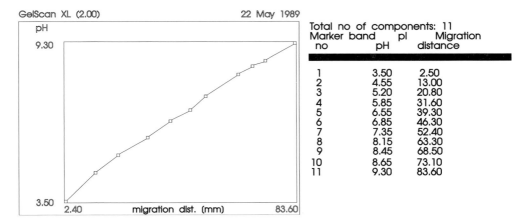

Total no of components: 11		
Marker band no	pI pH	Migration distance
1	3.50	2.50
2	4.55	13.00
3	5.20	20.80
4	5.85	31.60
5	6.55	39.30
6	6.85	46.30
7	7.35	52.40
8	8.15	63.30
9	8.45	68.50
10	8.65	73.10
11	9.30	83.60

Fig. 10: Calibration curve: pH values of pI markers in function of the position of the bands (pH gradient).

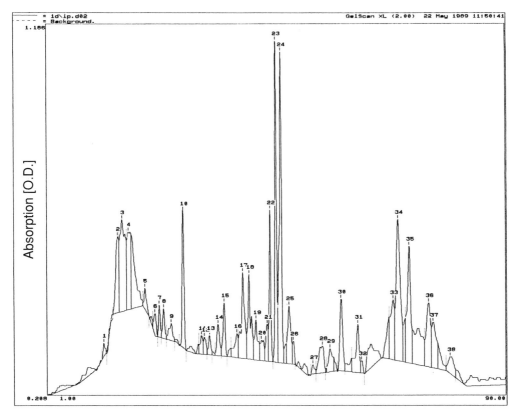

Fig. 11: Densitogram: IEF of proteins from pork extract.

No	Locn mm	Height AU	Area AU*mm	Rel ar %	pI pH
1	11.96	0.037	0.01708	0.6	4.44
2	14.68	0.150	0.10772	3.8	4.69
3	15.52	0.142	0.15056	5.3	4.76
4	16.76	0.035	0.02440	0.9	4.86
5	19.96	0.049	0.02384	0.8	5.13
6	21.88	0.038	0.01776	0.6	5.26
7	22.72	0.081	0.02912	1.0	5.31
8	23.56	0.075	0.03400	1.2	5.36
9	25.04	0.040	0.02988	1.1	5.45
10	27.40	0.360	0.15600	5.5	5.59
11	30.96	0.026	0.01100	0.4	5.81
12	31.56	0.025	0.01292	0.5	5.84
13	32.52	0.038	0.01492	0.5	5.93
14	34.20	0.077	0.04596	1.6	6.08
15	35.40	0.135	0.06676	2.4	6.19

Fig. 12: Classification of the pI values of the 15 most acidic proteins of Fig. 11 using the calibration line in Fig. 11.

Frequent problems:

More bands occur than given in the instructions. This is almost always the case when silver staining is used. What can be done?

The IEF marker is applied in two dilutions are:

   1) One dilution step for Coomassie staining: 1.5 µg/mL per protein, dilute the vial containing the lyophilized markers accordingly with double-distilled water.

   2) One dilution step in the concentration for the silver staining: approximately 1/50th of the first concentration.

Preparation (1) is applied on the edge of the gel and preparation (2) beside. Once the focusing is finished and before fixing, a so-called "isoelectric rod" is prepared; blue bands, which are unequivocal in this case, are marked with colored marker pens on a plastic ruler or a strip of paper. The anode and cathode should be marked with a black pen (Fig. 13).

   By comparing the focusing track, which has the concentration of preparation (2), with the "IEF rod", the bands, which should be used for identification according to the manufacturer, can be found among the many others.

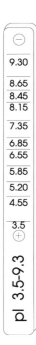

Fig. 13: "Isoelectric rod".

7 Perspectives

Electrode solutions: For applications, which require long separation times, basic gradients, and/or with presence of high molar urea, it is necessary to use filter paper strips with electrode solutions. In table 2 list of electrode solutions is given for 0.5 mm thin gels for the respective pH intervals.

Tab. 2: Electrode solutions for IEF in polyacrylamide gels.

pH Gradient	Anode	Cathode
3.5 – 9.5	0.5 mol/L H_3PO_4	0.5 mol/L NaOH
2.5 – 4.5	0.5 mol/L H_3PO_4	2% Ampholine pH5-7
2.5 – 4.5	0.5 mol/L H_3PO_4	0.4 mol/L HEPES
3.5 – 5.0	0.5 mol/L H_3PO_4	2% Ampholine pH6-8
4.0 – 5.0	0.5 mol/L H_3PO_4	1 mol/L glycine
4.0 – 6.5	0.5 mol/L acetic acid	0.5 mol/L NaOH
4.5 – 7.0	0.5 mol/L acetic acid	0.5 mol/L NaOH
5.0 – 6.5	0.5 mol/L acetic acid	0.5 mol/L NaOH
5.5 – 7.0	2% ampholine pH 4-6	0.5 mol/L NaOH
5.0 – 8.0	0.5 mol/L acetic acid	0.5 mol/L NaOH
6.0 – 8.5	2% ampholine pH 4-6	0.5 mol/L NaOH
7.8 – 10.0	2% ampholine pH 6-8	1 mol/L NaOH
8.5 – 11.0	0.2 mol/L histidine	1 mol/L NaOH

Rehydrated gels: A number of possible applications exist for rehydrated focusing gels. A few of these and the necessary rehydration solutions will be presented here.

The advantages of the gels have already been explained at the beginning.

Basic, water-soluble proteins: Basic carrier ampholytes partially inhibit the polymerization of polyacrylamide gels. In addition basic gradients are more prone to cathodic drift than others, partly because of acrylamide monomers and remains of APS in the gel. These problems do not arise when washed and rehydrated gels are used.

cathodic drift (Righetti and Drysdale, 1973)

Practice has shown, that electrode solutions must be applied to stabilize the gradient.

Proteins and enzymes sensitive to oxidation: 2-mercaptoethanol is sometimes used to prevent oxidation in the gel, but it inhibits polymerization. In addition ammonium persulfate possesses oxidative characteristics in the gel (Brewer, 1967). Once again, these problems do not occur with rehydrated gels.

Brewer JM. Science. 156 (1967) 256-257.

Heat-sensitive enzymes and enzyme-substrate complexes: 37% DMSO in the gel is generally used during cryo-isoelectric focusing (at −10 to −20 °C) (Righetti, 1977). In this case the gel is reconstituted with 37% DMSO (v/v) and 2% Ampholine pH > 5 (w/v).

Ampholytes with pH < 5 precipitate at these low temperatures.

Hydrophobic proteins: The non-ionic detergents required to solubilize hydrophobic proteins inhibit the co-polymerization of the gel and support film when they are added to the polymerization solution. Support free gels with non-ionic detergents bend during the run and can even tear at high field strengths. These problems do not occur with rehydrated gels: they can be reconstituted with PEG, Triton X-100, Nonidet NP-40 or zwitterionic detergents such as LDAO or CHAPS in any desired concentration.

Complex mixtures of proteins: If proteins from cell lysates or tissue extracts must be completely separated, the aggregation or precipitation of the proteins should be prevented.

Review article about these problems in relation to high resolution 2D electrophoresis:
Dunn MJ, Burghes AHM. Electrophoresis. 4 (1983) 97-116.

An 8 mol/L urea solution, 1 to 2% non-ionic detergent, 2.5% carrier ampholyte and about 1% 2-mercaptoethanol are usually used. During polymerization the problems mentioned can occur. In addition in these highly viscous solutions the mobility of the proteins is reduced, so the cathodic drift has more influence.

Separations in these highly viscous gels take a longer time then usual. Practice has shown, that electrode solutions must be applied to stabilize the gradient.

These problems disappear with the rehydration of already polymerized gels: washing out unwanted substances such as ammonium persulfate and acrylamide monomers reduces the cathodic drift.

Immunofixation: When IEF polyacrylamide gels with a *T* value of 5% or less have a thickness of 0.5 mm or less, antibodies are able to diffuse into the matrix. Thus, in some cases, TCA can be avoided for fixation, and the noninteresting proteins can be washed out of the matrix. This is particularly advantageous for the detection of oligoclonal IgG bands in serum and cerebrospinal fluid. After the nonprecipitated antibodies and proteins have been washed out, silver staining can be employed to detect the antigen-antibody complexes.

This detection method is only possible since thin and soft polyacrylamide gels on film-support have been introduced.

In order to save costs for antibodies, small gels are preferred for this technique.

Method 7: SDS-polyacrylamide electrophoresis

The principles of SDS-electrophoresis and its fields of application are described in the introduction.

see pages 26 and following

1 Sample preparation

The native state of proteins is shown in Fig. 1: the tertiary structure of a polypeptide coil – slightly simplified – and a protein (IgG) with a quaternary structure consisting of many subunits – highly simplified. This spatial arrangement is conditioned by intra- and intermolecular hydrogen bonds, hydrophobic interactions and disulfide bridges, which are formed between cysteine residues.

For SDS electrophoresis proteins must be converted from their native form into SDS-protein micelles.

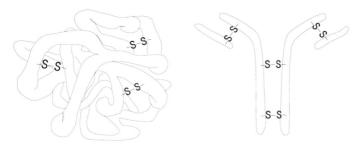

tertiary structure   quarternary structure

Fig. 1: Native structure of proteins.

SDS-treatment

The addition of an excess of SDS to protein solutions has the following effect:

● individual charge differences of the proteins are masked,

● hydrogen bonds are cleaved,

● hydrophobic interactions are canceled,

● aggregation of the proteins is prevented,

● in addition the polypeptides are unfolded (removal of the secondary structure) and ellipsoids are formed.

In the process 1.4 g of SDS are bound per gram of protein. All the micelles have a negative charge which is proportional to the mass.

see also pages 82 and following for qualitative and quantitative evaluation

The Stokes radii of the micelles are then proportional to the molecular weight (MW), and a separation according to the molecular weight is obtained by electrophoresis.

1 to 2% (w/v) SDS are used in the sample solution and 0.1% in the gel.

Some proteins, e.g. casein require 2 % SDS or more.

The method of sample treatment is very important for the quality of the separation and its reproducibility; the various possibilities will therefore be described in detail.

Stock buffer (pH 6.8):
    6.06 g of Tris
    0.4 g of SDS
    make up to 80 mL with double-distilled water
    titrate to pH 6.8 with 4 mol/L HCl
    make up to 100 mL with double-distilled water.

This in fact is the stock buffer for the stacking gel - and also for the sample buffer - in the original publication of Laemmli. This buffer has become a standard in SDS-electrophoresis.

For some proteins it is better to use a more basic buffer, e.g. Tris-HCl pH 8.8, because SDS binds better to proteins at alkaline pH.

Non-reducing SDS treatment

Many samples, e.g. physiological fluids such as serum or urine, are simply incubated with a 1% SDS buffer without reducing agent, because one does not want to destroy the quaternary structure of the immunoglobulins. The disulfide bonds are not cleaved by this treatment and so the protein is not fully unfolded (Fig. 2).

A proper molecular weight measurement is not possible here: for example albumin (68 kDa) has an apparent molecular weight of 54 kDa.

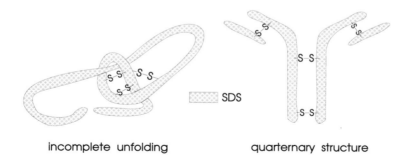

incomplete unfolding   quarternary structure

Fig. 2: Proteins treated with SDS without reducing agent.

Nonred SampB (Non reducing sample buffer):

EDTA is used to inhibit the oxidation of DTT in case the sample should be reduced. Dr. J. Heukeshoven, personal communication.

1.0 g of SDS	2.5 mL of stock buffer
3 mg of EDTA	make up to 100 mL with
10 mg of Bromophenol Blue	double-distilled water.

Diluting the samples (examples)

For *Coomassie staining* (sensitivity 0.1 to 0.3 mg per band):

Serum: 10 μL + 1.0 mL of sample buffer   *Apply 7 μL*

Urine: 1 mL + 10 mg of SDS (it should not be further diluted with sample buffer)   *Apply 20 μL*

For *silver staining* (sensitivity 1 to 30 ng per band):

Serum: dilute the above sample 1:20 with Nonred SampB.   *Apply 3 μL*

Urine: 1 mL + 10 mg of SDS (dilute 1:3 for some proteinurias).   *Apply 3 μL*

● Incubate for 30 min at room temperature, *do not boil!* Boiling non reduced proteins can lead to the formation of protein fragments.   *Protease activity!*

Reducing SDS treatment

Proteins are totally unfolded and a separation according to molecular weight is possible, only when a reducing agent like dithiothreitol (DTT) is added (Fig. 3). The smell is reduced by the use of the less volatile DTT instead of 2-mercaptoethanol.

Prepare DTT and Red SampB in the appropriate quantities shortly before use.

Another advantage of using DTT instead of 2-mercaptoethanol is , that when reduced and nonreduced sample should be separated in the same gel, DTT does not diffuse into the traces of non reduced proteins.

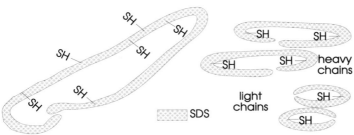

Fig. 3: SDS treated and reduced proteins.

Dithiothreitol stock solution (2.6 mol/L DTT):
dissolve 250 mg of DTT in 0.5 mL double-distilled water.

Red SampB (reducing sample buffer 26 mol/L):
10 mL of Nonred SampB + 100 μL of DTT solution

Add a bit of Orange G for better differentiation.

Diluting the samples (examples):

For Coomassie staining (sensitivity: 0.1 to 0.3 mg per band)
LMW marker + 415 μL of Red SampB   *Apply 5 μL, 1 mg of BSA per application*

HMW marker + 200 μL of double-distilled water, do not boil!   *Apply 5 μL*

CMW marker: LMW marker + 315 μL of Red SampB + 100 mL of desalted collagen.   *Apply 5 μL*

Meat extracts: 100 μL + 900 μL of H2O bidist, divide: *Apply 5 μL*
50 μL + 237.5 μL of Red SampB.

For *silver staining* (sensitivity 1 to 30 ng per band): dilute the above sample 1:20 with Red SampB.

Apply 3 μL

● Boil the samples for 3 min (heating block).

● After cooling:

Add 1 μL DTT solution per 100 μL of sample solution. The reducing agent can be oxidized during heating. The SH groups are better protected by the addition of reducing agent which prevents refolding and aggregation of the subunits.

The oxidation of DTT is inhibited by EDTA , so this treatment is not necessary for all samples: it is best to perform a comparison test.

In practice the (repeated) renewed addition of reducing agent is often forgotten: this results in additional bands in the high molecular weight area ("ghost bands") and precipitation at the application point.

Reducing SDS treatment with alkylation

The SH groups are better and more durably protected by ensuing alkylation with iodoacetamide (Fig. 4). Sharper bands result; in proteins containing many amino acids with sulfur groups a slight increase in the molecular weight can be observed.

In addition, the appearance of artifact lines during silver staining is prevented because iodoacetamide traps the excess DTT. Alkylation with iodoacetamide works best at pH 8.0 which is why another sample buffer with a higher ionic strength (0.4 mol/L) is used. This high molarity is not a problem for small sample volumes. Dr. J. Heukeshoven, personal communication.

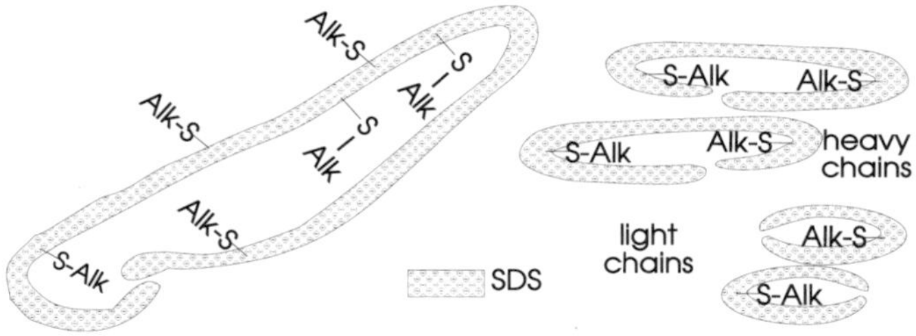

Fig. 4: Reduced and alkylated proteins treated with SDS.

Sample buffer for alkylation
(pH 8.0: 0.4 mol/L):

4.84 g Tris
1.0 g SDS
3 mg EDTA → 80 mL with H_2O_{bidist},
titrate to pH 8.0 with 4 mol/L HCl,
make up to 100 mL with H_2O_{bidist},
add 10 mg of Bromophenol or Orange G,
10 mL of sample buffer (Alk) + 100 μL of DTT solution.

*Iodoacetamide solution
20% (w/v):*
20 mg iodoacetamide
+ 100 μL H_2O_{bidist} .

This is the limit of saturation for iodoacetamide! The percentile weight error is negligible. The additional dilution of the sample by the alkylation solution can be taken into account either during the sample preparation or the application: apply 10% more sample volume.

After boiling the reduced sample:

Add 10 μL of iodoacetamide solution per 100 μL of sample solution and incubate for 30 min at room temperature.

To quantify the protein bands it is recommended that an external standard in the form of a series of dilutions of marker proteins is separated in the same gel (Tab. 1). Different staining efficiencies can thus be taken into account.

Tab. 1: Dilution series of marker proteins for an external standard for quantification by densitometry.

Quantification standard	Quantity applied	
	μL	μg
LMW-Marker + 200 μL sample buffer	10	4.15
LMW-Marker + 200 μL sample buffer	7	2.90
LMW-Marker + 200 μL sample buffer	5	2.01
LMW-Marker + 400 μL sample buffer	7	1.45
LMW-Marker + 600 μL sample buffer	5	0.69
LMW-Marker + 1000 μL sample buffer	5	0.42

2 Stock solutions for gel preparation

Acrylamide, Bis solution "SEP" (T = 30%, C = 2%):
29.4 g of acrylamide + 0.6 g of bisacrylamide, make up to 100 mL with double-distilled water (H_2O_{Bidist}).

*C = 2% in the gradient gel solution prevents the **separation** gel from peeling off the support film and cracking during drying.*

Acrylamide, Bis solution "PLAT" (T = 30%, C= 3%):
Add 100 mL of H_2O_{Bidist} to the PrePAG mix (29.1:0.9).
Caution! *Acrylamide and bisacrylamide are toxic in the monomeric form. Avoid skin contact and dispose of the remains ecologically (polymerize the remains with an excess of APS).*

*This solution is used for slightly concentrated **plateaus** with C = 3%, because the slot would become unstable if the degree of polymerization were lower.*

Gel buffer pH 8.8 (4 × conc):
18.18 g of Tris + 0.4 g of SDS, make up to 80 mL with H_2O_{dist}. *Titrate to pH 8.8 with 4 mol/L HCl; make up to 100 mL with H_2O_{dist}.*

Ammonium persulfate solution (APS):
Dissolve 400 mg of APS in 1 mL of H_2O_{dist}.

Can be stored for one week in the refrigerator (4 °C).

Cathode buffer (conc):
7.6 g of Tris + 36 g of glycine + 2.5 g of SDS, make up to 250 mL with H_2O_{dist}.

Do not titrate with HCl!

Anode buffer (conc):
7.6 g of Tris + 200 mL of H_2O_{dist}. Titrate to pH = 8.4 with 4 mol/L HCl; make up to 250 mL with H_2O_{dist}.

Economy measure: the cathode buffer can also be used here.

3 Preparing the casting cassette

Gels with a completely smooth surface can be used in horizontal SDS-electrophoresis for sample application methods similar to IEF. When the gels are hand-made, it is possible to polymerize sample wells in the surface of the gel. For this a "slot former" is made out of the spacer:

The "spacer" is the glass plate with the 0.5 mm thick silicone rubber gasket glued on to it.

Preparing the slot former

To make sample application wells, a mould must be fixed on to the glass plate. The cleaned and degreased spacer plate is placed on the template ("slot former" template in the appendix) on the work surface. A layer of "Dymo" tape (embossing tape, 250 μm thick) is placed on the area which will be used as starting point avoiding air bubbles. The slot former is cut out with a scalpel (Fig. 5). After pressing the individual sample wells once again against the glass plate the remains of tape are removed with methanol. Several superimposed layers of "Scotch tape" quality (one layer = 50 μm) can be used instead .

"Dymo" tape with a smooth adhesive surface should be used. Small air bubbles can be enclosed when the adhesive surface is structured, these inhibit polymerization and holes appear around the slots.

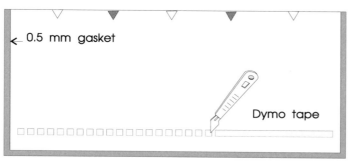

Fig. 5: Preparing the slot former.

In some applications, the sample must be applied over the entire gel width for subsequent blotting for producing test strips. In these cases the "Dymo" tape is glued over the entire width, and only cut in the center for the application of markers (see method 8 Blotting)

The next step is to make the casting cassette hydrophobic. This is done with a few mL of Repel Silane which are spread over the whole slot former with a tissue. This operation should be carried out under the fume hood. When the Repel Silane is dry, the chloride ions which result from the coating are rinsed off with water.

Assembling the casting cassette

The gel is polymerized on a support film for mechanical support and easier handling. A glass plate is placed on an absorbent tissue and wetted with a few mL of water. The GelBond PAG film is applied with a roller, the untreated, hydrophobic side down (Fig. 6). This creates a thin film of water between the support film and the glass plate which holds them together by adhesion.

The excess water which runs out is soaked up by the tissue. To facilitate pouring the gel solution, the film should overlap the long edge of the glass plate by about 1 mm.

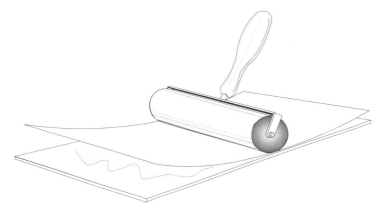

Fig. 6: Applying the support film with a roller.

The slot former is placed over the glass plate and the cassette is clamped together (Fig. 7).

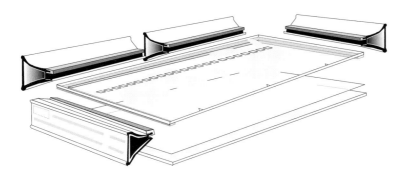

Fig. 7: Assembling the casting cassette.

The casting cassette is cooled to 4 °C in the refrigerator for about 10 min; this delays the onset of polymerization. This last step is essential, since it takes the poured gradient about 5 to 10 min to settle horizontally.

In a warm laboratory - during the summer for example - the gel solution should also be cooled to 4 °C.

4 Gradient gel

Preparation of the three gel solutions for a gradient T=8% to 20% and a sample application plateau with T=5% (Tab. 2).

A "super dense" plateau does not mix with the gradient.

Tab. 2: Composition of the gel solutions.

APS s. p. 177

Pipette into 3 test tubes	Super dense $T = 5\%, C = 3\%$	Dense $T = 8\%, C = 2\%$	Light $T = 20\%, C = 2\%$
Glycerol (87 %)	6.5 mL	4.3 mL	–
Acrylamide, Bis"PLAT"	2.5 mL	–	–
Acrylamide, Bis"SEP"	–	4.0 mL	10 mL
Gel buffer	3.5 mL	3.75 mL	3.75 mL
Bromophenol Blue	–	–	100 µL
Orange G	100 µL	–	–
Temed	10 µL	10 µL	10 µL
bring to final volume with double-distilled water	*15 mL*	*15 mL*	*15 mL*

Pouring the gradient

a) Assembling the casting cassette

The gradient is prepared with a *gradient mixer*. It is made of two communicating cylinders. The front cylinder, the *mixing chamber*, contains the denser solution and a magnetic stirrer bar. The back cylinder, the *reservoir*, contains the lighter solution. The terms "dense" and "light" demonstrate that a *density gradient* is coupled to the acrylamide solution: the dense solution contains about 25% glycerol and the light one 0%. The difference in densities prevents the solutions from mixing in the cassette and allows the gradient to settle horizontally.

It is not recommended to use sucrose for ultrathin gels because the solution would be too viscous.

Because of the difference in densities, opening the channel between the two chambers would cause the denser solution to flow back into the reservoir.

The *compensating bar* in the reservoir corrects for the difference in density and for the volume of the magnetic stirrer.

In the gradient presented here, the dense solution contains the lower proportion of acrylamide and gives the part of the gel with the larger pores. The light solution contains more acrylamide and yields the part with the smaller pores. In consequence the slot former is placed in the lower part of the cassette (Fig. 8 and 9).

This is in contrast to the conventional gradient pouring technique for verrtical chambers, but has a number of advantages:

- The part with the small pores keeps its pore size without glycerol.

  During separation, the part containing glycerol slowly swell because glycerol is hygroscopic.

- A high proportion of glycerol in the sample application area prevents the gel from drying out, increases the stability of the sample slots, improves the solubility of high molecular peptides and compensates for the effect of the salts in the sample.

  In many cases dialysis of the sample can be avoided.

- Only one acrylamide stock solution is needed for a gel concentration of up to T = 22.5%.

  When T = 30% it does not crystallize in the refrigerator.

- Because the highly concentrated acrylamide solution is on top, the settling of the gradient is not disturbed by thermal convection.

  The higher the acrylamide concentration, the more heat is released during polymerization.

- In the part of the gradient with the small pores, where the influence of the matrix on the zones in the most important, the perfect levelling of the gradient is important. Glycerol free solutions are less viscous.

  The viscosity of the solution plays an important role for ultrathin gels.

To pour a linear gradient (Fig. 8) both cylinders of the gradient mixer are left open. The laboratory platform ("Laborboy") is set so that the outlet lies 5 cm above the upper edge of the gel.

For reproducible gradients the outlet of the gradient mixer must always be on the same level above the edge of the gel cassette.

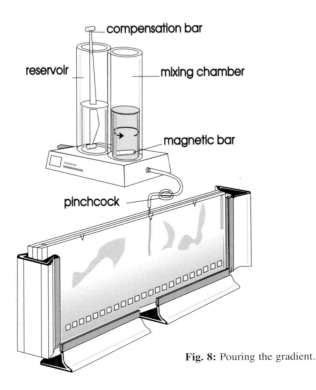

Fig. 8: Pouring the gradient.

Before filling the connecting channel between the *reservoir* and the *mixing chamber* as well as the *pinchcock* are shut.

The stirring bar is then placed in the mixing chamber and the optimum speed set on the magnet stirrer.

b) Casting the gel

3.5 mL are needed for the plateau. Since not all 15 mL of the very dense solution must be polymerized, a fourth vial containing gel solution is placed alongside the three prepared (cooled) vials so as to put aside the 3.5 mL of solution for the plateau (Fig. 9).

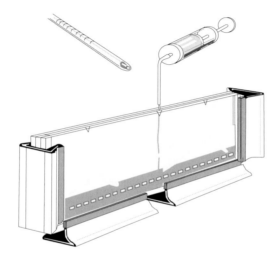

Fig. 9: Casting the gel solution for the plateau.

The dense and the light solutions are poured directly into the gradient mixer in the following steps:

● pour the light solution into the reservoir,

● briefly open the valve to fill the connecting channel,

● pour the dense solution into the mixing chamber.

When this is done, the cassette can be removed from the refrigerator and connected to the gradient mixer with the slot former oriented towards the gradient mixer.

The glass plate with the film is oriented towards the user.

The following operations should be carried out to fill the cassette:

● Add APS to the plateau solution set aside (Tab. 3).

Tab. 3: Catalysts volumes and concentrations.

Gel solution	Volume mL	APS μL
Very dense	3.5	5.0
Dense	7.0	6.2
Light	7.0	4.0

● Pipette the 3.5 mL of the very dense solution into the cassette (Fig. 9);

The notches of the slot former are oriented towards the gradient mixer.

● pipette APS into the reservoir;

● disperse APS while introducing the compensating stick into the reservoir;

● pipette the APS into the mixing chamber and stir briefly but vigorously with the magnetic stirrer;

to disperse the catalyst

● place the outlet in the middle notch of the slot former;

● set the magnetic stirrer at moderate speed;

do not generate air bubbles

● open the outlet valve;

● open the connecting valve.

The gradient mixer should be empty when the fluid level has reached the top of the cassette.

Rinse out the mixer with double-distilled water immediately afterwards.

100 μL of 60 % (v/v) isopropanol-water are then layered in each filling notch. Isopropanol prevents oxygen, which inhibits polymerization, from diffusing into the gel. The gel will then present a well-defined, aesthetic upper edge.

The gradient which is still irregular because of the convection movement caused by pouring the solution takes about 10 min to level out before polymerization starts. The gel should be solid after about 20 min.

The gel must be completely polymerized before electrophoresis, because the electrophoretic mobilities of the protein-SDS micells are greatly influenced by the sieving properties of the gel.

At least for 3 hours, ideally overnight, at room temperature.

Although, when the gel seems to have perfectly polymerized, it can not be used immediately. The matrix only becomes regular after a slow "silent polymerization". If it is used too early, the bands are shaky, not straight and even as usual.

5 Electrophoresis

Preparing the separation chamber

- Turn on the cooling system: + 15 °C;
- Unclamp the gel cassette and place the gel with the slot former on the bottom on the cooling plate of the Multiphor;
- Place the cassette vertically to ease the glass plate from the GelBond film with a thin spatula;
- Grasp the film at a corner where the acrylamide concentration is high and pull it away from the slot former.

Placing the gel on the cooling plate

Wet the cooling plate with 1 mL of kerosene or DC-200 silicone oil.

Water is not suitable as it can cause electric shorting.

- Place the gel onto the center of the cooling plate with the film on the bottom; the side with the slots must be oriented towards the cathode (−). Avoid air bubbles.

- Shift the gel so that the anodal edge coincides exactly with line "5" on the scale on the cooling plate (Fig. 11).

wear gloves

- Lay two of the electrode wicks into the compartments of the PaperPool. If smaller gel portions are used, cut them to size.

*Be sure to use **very** clean wicks, SDS would dissolve any traces of contaminating compounds.*

- Mix 10 mL of the cathode buffer with 10 mL distilled water and apply it onto the respective strip (Fig.10).

less volume for shorter strips

- Mix 10 mL of the anode buffer with 10 mL distilled water and apply it onto the respective strip.

less volume for shorter strips

- Place the cathode wick onto the cathodal edge of the gel; the edge of the wick matching "3.5" on the cooling plate; the anodal wick over the anodal edge, matching "13.5".

Always apply cathode wick first to avoid contamination of catho-de buffer with leading ions.

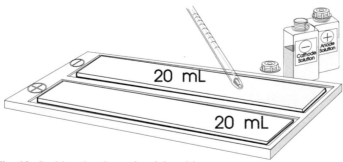

Fig. 10: Soaking the electrode wicks with the respective solutions in the PaperPool.

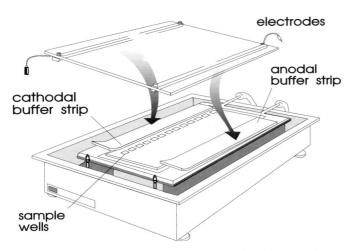

Fig. 11: Appliance for semi-dry SDS electrophoresis with electrode wicks.

Sample application

● Rapidly pipette the samples in the slots.

If a gel with a smooth surface, that is without sample wells, is used, the side of the gel with the large pores should be marked. The gel is then, as described above, placed on the cooling plate so that the samples are also applied on line "5".

Ready-made gels are marked by cutting off a corner of the film on the side with narrow pores .

The side with large pores must be placed towards the cathode!

The samples are applied 1 cm from the cathode strips and at least 1.5 cm from the sides. There are several possibilities of application (Fig. 12):

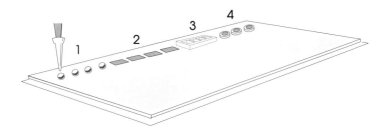

Fig. 12: Possibilities of sample application on SDS gels with a smooth surface.

● Directly pipette droplets (1).

It is recommended not to use more than 3 μL.

● Place the sample application pieces on the gel and then apply the sample (2).

It is possible to apply 3 to 20 μL.

● Use a sample application strip and pipette the samples in the slots (3).

It is possible to apply 30 to 40 µL.

● Remove some Raschigg rings from the condenser of a rotatory evaporator, place them on the gel, apply the samples (4).

It is possible to apply up to 100 µL.

Electrophoresis

● Clean platinum electrode wires before (and after) each electrophoresis run with a wet tissue paper. Move electrodes so that they will rest on the outer edges of the electrode wicks.

The buffer ions must be in between the electrodes.

● Connect the cables of the electrodes to the apparatus and lower the electrode holder plate (Fig. 11).

Be sure, that the electrodes have complete contact on the wicks.

● Close the safety lid.

Separation conditions:
1000 V, 50 mA, 35 W, 1 h 30 min.

with normal power supply

Tab. 4: Program for a programable power supply.

Phase	U V	I mA	P W	t h:min	*) Vh
1	200	50	30		85
2	600	50	22	1:20	
3	100	5	5	1:00	

* U/t-integral
Phase 1 is for a gentle sample entry, phase 3 works against band diffusion during unsupervised electrophoresis.

● Switch off the power supply;
● Open the safety lid;
● Remove the electrode strips from the gel and dispose them;
● Remove the gel from the cooling plate.

6 Coomassie and silver staining

Quick Coomassie staining

● *Staining:* hot staining by stirring with a magnetic stirring bar in a stainless steel tank: 0.02% Coomassie R-350: Dissolve 1 PhastGel Blue tablet in 1.6 L of 10 % acetic acid; 15 min, 50 °C (see Fig. 13);

The gel is placed face down on the grid.

● Destaining: in the MultiWash in 10% acetic acid for 2 h at room temperature (see Fig. 14);

The MultiWash is composed of a destaining tank, a pump and active charcoal mixed bed ion-exchange cartridges.

- Preserving: in a solution of 25 mL of glycerol (87% w/v) + 225 mL of distilled water for 30 min;
- *Drying:* air-dry (room temperature).

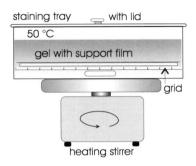

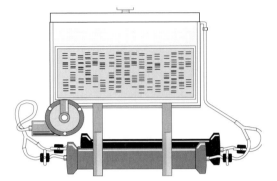

Fig. 13: Device for hot staining.

Fig. 14: MultiWash. Destaining and gel washing device.

Colloidal staining

This method has a high sensitivity (ca. 30 ng per band), but takes overnight. There is no background destaining (Neuhoff *et al.* 1985).

Neuhoff V, Stamm R, Eibl H. Electrophoresis. 6 (1985) 427-448.

Preparation of the staining solution:

Slowly add 100 g of ammonium persulfate to 980 mL of a 2% H_3PO_4 solution till it has completely dissolved. Then add Coomassie G-250 solution (1 g per 20 mL of water). Do not filter the solution! Shake before use.

The staining solution can be used several times.

Staining

- Fixing 1 h in 12% (w/v) TCA, for gels on support films it is best to place the gel surface on the bottom (on the grid of the staining tank), so that additives with a higher density (e.g. the glycerol for gradient gels) can diffuse out of the gel.

The use of a stirrer is recommended for this step.

- Staining overnight with 160 mL of staining solution [0.1% (w/v) Coomassie G-250 in 2% H_3PO_4, 10% (w/v) ammonium persulfate see above] plus 40 mL of methanol (add during staining).

- *Wash* for 1 to 3 min in 0.1 mol/L Tris, H_3PO_4 buffer, pH 6.5.

- *Rinse* (max. 1 min) in 25% (v/v) aqueous methanol.

- *Stabilize* the protein-dye complex in 20% aqueous ammonium sulfate.

Silver staining

Tab. 5: Silver staining acc. to Heukeshoven and Dernick (1986).

Step	Solution	V mL	t min
Fixing	300 mL ethanol 100 mL acetic acid with $H_2O_{dist} \rightarrow$ *1000 mL*	250	>30
Incubation	75 mL ethanol*) 17.00 g Na-acetate 1.25 mL glutaraldehyde (25% w/v) 0.50 g $Na_2S_2O_3 \times 5\ H_2O$ with $H_2O_{dist} \rightarrow 250$ mL	250	30 or overnight
Washing	H_2O_{dist}	3×250	3×5
Silvering	0.5 g $AgNO_3$**) 50 µL formaldehyde (37% g/v) with H_2Odist $\rightarrow 250$ mL	250	20
Development	7.5 g Na_2CO_3 30 µL formaldehyde (37% w/v) with $H_2O_{dist} \rightarrow 300$ mL if pH >11,5, titrate to this value with $NaHCO_3$ solution	1×100 1×200	1 3 to 7
Stopping	2.5 g glycine with $H_2O_{dist} \rightarrow 250$ mL	250	10
Washing	H_2O_{dist}	3×250	3×5
Preserving	25 mL glycerol (87% g/v) with H_2Odist $\rightarrow 250$ mL	250	30
Drying	air-drying (room temperature)		

* First dissolve NaAc in water, then add ethanol. Add the thiosulfate and glutaraldehyde just before use.
** Dissolve $AgNO_3$ in water, add the formaldehyde before use.

Blue Toning
   In general, silver stained bands can not easily be quantified, because the staining curve is very steep. Frequently the bands show different colors; highly concentrated fractions show hollow bands ore bands with yellow centers. These images severely

interfere with semiquantitative and qualitative evaluations using densitometers or scanners.

The evaluation of these results can be improved by "blue toning" according to Berson (1983): After silver staining the gel must be washed thoroughly with double-distilled water and then it is immersed for 2 minutes in a freshly mixed bath of:

Berson G. Anal Biochem. 134 (1983) 230-234.

This process is adopted from the photoshops to make blue slides out of photo negatives.

140 mL H_2Odist + 20 mL of 5% $FeCl_3$ + 20 mL of 3 % oxalic acid + 20 mL of 3.5 % potassium hexacyanoferrate.

Blue toning is slightly improving the sensitivity and gives uniformly stained bands.

Place the gel in water and then in glycerol solution before drying.

7 Blotting

The specific detection of proteins can be carried out after their electrophoretic transfer from the gel to an immobilizing membrane (blotting membrane). Either different samples are separated on the gel and the membrane is analysed with one antibody solution, or one antigen is loaded over the entire gel width and the membrane is cut into strips for probing in different patient sera.

To blot gels which have been stained the proteins must be solubilized in SDS buffer again (by soaking the gels in the SDS buffer). In most cases, for subsequent immuno-detection, the antigen-antibody reactivity remains despite additional denaturation with staining reagents (Jackson and Thompson, 1984). If blotting is carried out immediately after electrophoresis or with handmade gels and when the entire gel must be blotted, the gel is poured on the hydrophobic side of a GelBond PAG film from which it can easily be removed after electrophoresis.

Jackson P, Thompson RJ. Electrophoresis. 5 (1984) 35-42.

see method 8, pages 189 and following

8 Densitometry

a) Laser Densitometry

This is the most exact quantitative evaluation of an SDS gel, stained with colloidal Coomassie Blue. The laser densitometer data are processed with the appropriate software on a personal computer.

The *densitometer* is programmed in the menu as follows: In "Plot-Menu" autohold on "no", In "Setup-Menu" on "1D-Mode" and set "Shape of Beam" = "Line".

The interface RS 232 must be chosen as "Type of output".

It is best if the gel always lies in the same orientation and place on the measuring tray of the laser densitometer. The end of the gel with the sample application wells is placed towards the front of the light box so that the side of the slots in the separation direction lie in position 20 on the Y axis (Fig. 15).

This leads to increased possibilities of comparison between different gels and facilitates evaluation.

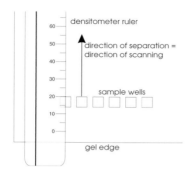

Fig. 15: Placing the gel on the light box of the densitometer to determine the exact molecular weight.

The entry of the measurement parameter is done in the "Define Track" menu. So that all 7 mm of the electrophoresis track are covered, it is recommended to enter "X-WIDTH" = 8 (6.4 mm).

The "X-WIDTH" indicates how often the laser beam scans the sample trail shifting of 800 μm each time.

The resolution of the densitometer is defined with "Y-STEP". A highly resolved gel electrophoresis should be scanned with the same resolution, that is with "Y-STEP" = 1.

The motor stops every 40 μm when "Y-STEP" = 1. The resolution can be chosen from 20 μm to 600 μm.

The electrophoretic separation starts at the edge of the sample application well which lies at position 20 on the light box, therefore: Y-START "20". The separation distance is 100 mm: Y-STOP "120".

All tracks are marked with the ruler by pressing the direction arrows so that the support line on the ruler lies over the middle of the track. The position of the track is saved with the command "SAVE". Leave the "Define Track" mode with "ESC".

On the *computer*, the program is loaded. Choose the "Scan" program in the "Main Menu". The name of the "Datafile" can be chosen at will, e.g.. "SDS-Gel". If several gels have the same name, choose the corresponding number.

After entering the last comment, the computer executes a communication test with the densitometer. The measurement of all the preprogrammed tracks and the transmission of all the data to the computer is done automatically.

Assignment of the molecular weights

To check the edited method, it is recommended to load the track of the molecular weight marker and to evaluate it. The molecular weights yielded by the program must correspond to the marker protein peaks.

Quantification of the bands with an external standard

For quantifying *known* proteins, it is very important to have calibration standards for densitometry as for the quantification of chromatographic separations. The protein to be analyzed must be pure, a known quantity is applied on the electrophoresis gel and is used as standard during the separation. It is recommended to apply different quantities of external standard in each case, since almost no substances show a linear concentration/extinction relationship. The use of different calibration points permits these non-linear functions to be taken into account (see Fig. 52 on page 84).

For quantification of *unknown or impure* proteins, "albumin equivalents" are used, albumin being the external standard. Naturally, it is not possible to obtain an absolute quantification with this method. But the use of a definite protein, whose peaks are compared with the peaks of all the other bands, gives a good relative value of the amount of an unknown protein. The comparison of different gels is guaranteed by the standard amount of albumin used each time.

see page 91

An example of the different dye affinities of various proteins is described in section 1.

Once the integration is finished, the results can be printed .The data can also be processed with new developed *softwares*, as some of them are compatible with the laser densitometer.

see figure 56, page 89

b) Desk Top Scanner

Scanning with a desk top scanner is much easier and faster, and all parameters are controlled by the computer software with a graphical interface. Dialogue boxes ask for the necessary informations; many functions are controlled with the mouse

The gel can be laid onto the glass plate of the scanner at any place. The area of interest is mostly defined after scanning during image processing. After defining the tracks, the bands are detected automatically and marked on the display.

Background subtraction is performed by defining an empty background stripe with the mouse.

Quantification of the bands is then performed automatically with preset parameters: sensitivity, minimum density, noise filter, and shoulder sensitivity. Each band can be approached with the mouse, known data about the band are displayed on the screen. New softwares have a corrective function for adjusting tracks, which are not perfectly straight, and can match comparable bands in different tracks. The relative quantities of certain bands are then compared across all tracks of the gel, and they are displayed as histograms (see Fig. 16). Also bands in different gels can be compared with each other.

When known amounts of proteins are run in the gel, the amount of unknown samples of the same proteins can be calculated with the help of a calibration curve.

When one track contains marker proteins, the molecular weights of the separated proteins can be calculated as well.

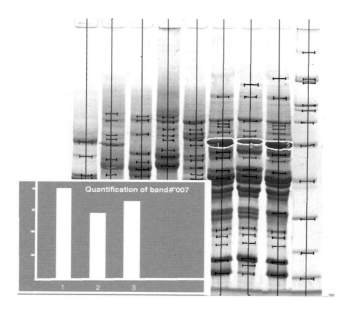

Fig. 16: Automated band detection with histogram presentation of relative amounts of proteins in a certain band across the tracks.

The band patterns can be displayed and printed as pherograms, densitometry curves, and tables. The data can be exported into desk top and spreadsheet programs, the images converted into a TIFF formate.

9 Perspectives

Most problems can be solved with the methods for SDS electrophoresis presented here. Should difficulties occur nevertheless, refer the trouble-shooting guide in part 3. A few other conceivable problems are described in part 1.

see pages 31 and following

Gel characteristics: The gel described here has an acrylamide gradient from $T = 8$ to 20% and a sample application plateau with $T = 5\%$. If a higher resolution in a narrower molecular weight range is required, a flatter gradient (e.g. $T = 10$ to 15%) or a homogeneous resolving gel (e.g. $T = 10\%$) are used.
For $T = i\,\%$, the volumes of acrylamide-Bis solution used for 15 mL of polymerization solution can easily be calculated as follows:

This also enables omplete separation of complex protein mixtures with a molecular weight range of 5 to 400 kDa.

$$V \text{ mL} = i \times 0.5 \text{ mL}$$

Different gel compositions will influence the separation time. For flatter gradients and homogenous gels, it is advisable to end the separation when the front has reached cathode.

Front is visualized by Orange G or Bromophenol Blue marker dyes.

SDS electrophoresis in washed and rehydrated gels

A series of experiments with rehydrating washed and dried gels – like in method 4 of this book – for SDS electrophoresis has shown, that the standard Tris-HCl / Tris-glycine buffer system can not be applied for these gels: the separation quality is very poor.

Obviously, this recipe can only work in presence of APS, Temed and monomers of acrylamide and Bis.

Good results are, however, are obtained with the Tris-acetate / Tris-tricine buffer system (as shown in Fig. 18 A):

The technique is almost only feasible with plastic-film-supported gels.

Rehydration buffer: 0.3 mol/L Tris / acetate pH 8.0, 0.1% SDS.
Cathode buffer: 0.8 mol/L Tricine, 0.08 mol/L Tris, 0.1 % SDS.
Anode buffer: 0.6 mol/L Tris / acetate pH 8.4, 0.1 % SDS.

20 mL in wick
20 mL in wick

SDS Disc electrophoresis in a rehydrated and selectively equilibrated gel

For the separation of samples with very high protein concentrations in one fraction (e.g. in pharmaceutical quality control) and / or very complex protein mixtures, complete separations of all fractions is only achieved, when all four discontinuities of disc electrophoresis are applied (see page 28 of this book). In conventional ready made gels, only the discontinuities in the gel matrix and between the leading ion in the gel and the trailing ion in the cathodal buffer can be applied.

An efficient stacking of the proteins in the first phase of electrophoresis is necessary here.

When a gel is supported by a rigid plastic film, its stacking zone can selectively be equilibrated in a stacking gel buffer with a different pH and a lower buffer concentration, short before use. This is performed in a vertical equilibration chamber (Fig. 17).

For a 0.5 mm thin gel equilibration takes 15 min.

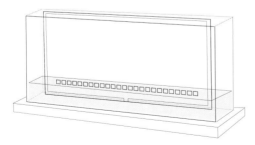

Fig. 17: Selective equilibration of the stacking gel zone in the appropriate buffer for disc electrophoresesis.

If a laboratory made gel is used with Tris-HCl pH 8.8 – like described in this chapter – , and Tris-glycine is used in the cathode buffer, best results are obtained with a stacking buffer containing 1.25 mol/L Tris-HCl pH 6.7 and 0.1 % SDS.

For a Tris-acetate buffer pH 8.0 and tricine in the cathode , use a stacking buffer containing 0.1 mol/L Tris-acetate pH 5.6 and 0.1 % SDS.

Peptide separation

Very good separations of low molecular weight peptides are obtained in SDS polyacrylamide electrophoresis with gels of relatively high acrylamide concentration, and employing a buffer with pH 8.4 (the pK-value of the basic group of the tricine, which is used in the cathode buffer) as well as high molarity of Tris according to Schägger and von Jagow (1987). Pharmacia had introduced a PhastGel® High Density, a 20 % T gel containing 30 % ethylenglycol for peptide separations (Pharmacia, 1987).

see page 34

Pharmacia TF 112: SDS-PAGE of low molecular weight proteins using PhastGel high density (1987).

By rehydration of a washed and dried gel with T= 15 % in a SDS buffer with 0.7 mol/L Tris-acetate pH 8.4 and 30 % ethylenglycol, a gel with very high resolving power in the low molecular weight area (1 to 40 kDa) is obtained within one hour. This gel should be run with the Tris-acetate anode buffer and the Tris-tricine cathode buffer according to the recipes given above. Electrophoresis takes 2 hours 45 minutes.

A separation result obtained in such a gel is shown in figure 18B.

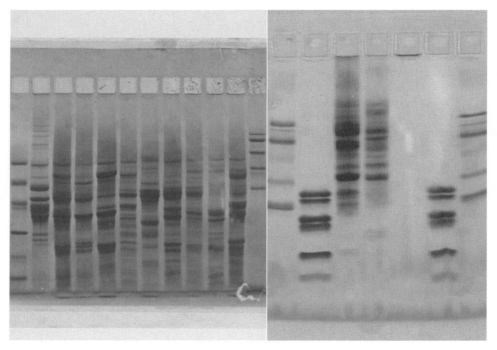

Fig. 18: A. SDS electrophoresis of legume seed extracts and molecular weight standards in a rehydrated discontinuous gel with a 5 % T stacking zone and a 10 = % T separating zone. B. Separation of peptide and molecular weight markers in a rehydrated 15 % T gel containing 30 % (v/v) ethylenglycol and the buffer described above. Staining with Coomassie Brilliant Blue R-350.

Method 8: Semi-dry blotting of proteins

The principles as well as a few possibilities of application are described in section I [1]). This chapter concerns the transfer technique as well as selected staining methods for blotting from SDS gradient pore gels according to method 7 [2]) and 15 [3]), and from IEF gels according to method 6 [4]). For the electrotransfer, the cooling plate is taken out of the chamber and the graphite plates are used instead (Fig. 1).

[1] *see pages 59 and following*
[2] *see pages 165 and following*
[3] *see pages and following*
[4] *see pages 151 and following*

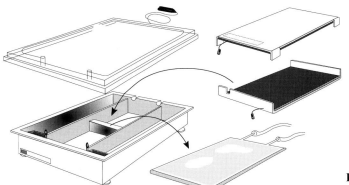

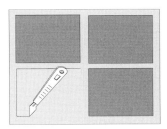

Fig. 2: Mask.

Fig. 1: Installing the graphite anode in place of the cooling plate.

A few practical tips will be given first:

- The filter paper and the blotting membrane must be cut to the size of the gel so that the current does not flow around the sides of the actual blot sandwich. Small blot sandwiches can be placed beside one another.

  The largest gel size is 20×27 cm. When a gel of the size described in method 7 should be blotted, it is easier to cut the gel into two halves before blotting.

- The filter paper does not have to be cut if a plastic frame is made in which openings for the gels and blotting membrane are cut out. This frame is placed on the stack of anodic filter paper before the blot is built up further (Fig. 2).

- Only gels of the same kind (IEF, SDS) can be blotted together because the transfer time depends on the pore size and the state of the proteins.

  Unfolded SDS-polypeptide micelles are different from focused proteins in globular form.

- The transfer buffer contains 20% methanol, so that the gels do not swell during the transfer and the binding capacity of the nitrocellulose is increased.

  If the biological activity of enzymes must be preserved, the methanol should be omitted.

● When gels are bound to a plastic film, the gel and the support film should first be separated with the Film Remover.

A thin wire is pulled between the gel and support film.

● When several gel-blot layers are blotted over one another, a dialysis membrane (Cellophane) is placed between each layer so that proteins which might migrate through, do not reach the next transfer unit.

Not recommeded, because when several transfer units are blotted, transfer efficiency is lost in the direction of the cathode.

● For continuous buffer systems, the same buffer with a pH of 9.5 is used for both the anodic and cathodic sides of the blots. To improve the transfer of hydrophobic proteins and to charge focused proteins (proteins are not charged at their pI), the buffer also contains SDS.

This method is the easiest. But the transfer is not as regular, and the bands not as sharp as in discontinuous buffer systems.

In discontinuous buffer systems the speed of migration of the proteins changes during the transfer because of the different ionic strengths of both the anode buffers (0.3 mol/L and 0.025 mol/L). This means that fewer proteins are transferred. The slow terminating ion in the cathode buffer compensates for the differences in speed of migration of the leading ions (here the proteins): a more regular transfer is obtained. In this case, SDS is only added in the cathode buffer.

This method is especially valuable for native blotting. The small amounts of SDS (0.01%) in the cathode buffer do not denature the proteins during the short contact time.

1 Transfer buffers

Continuous buffer system:

39 mmol/L glycine	2.930 g,
48 mmol/L Tris	5.810 g,
0.0375% (w/v) SDS	0.375 g,
20% (v/v) methanol	200 mL,

make up to 1000 mL with distilled water.

Discontinuous buffer system:

acc. to Kyhse- Andersen (1984)

Anode solution I:

0.3 mol/L Tris	36.3 g,
20% (v/v) methanol	200 mL,

make up to 1000 mL with distilled water.

Anode solution II:

25 mmol/L Tris	3.03 g,
20 %(v/v) methanol	200 mL,

make up to 1000 mL with distilled water.

Cathode solution

40 mmol/L 6-aminohexanoic acid	5.2g,
0.01% (w/v) SDS	0.1 g,
20% (v/v) methanol	200 mL,

make up to 1000 mL with distilled water.

The same as ε-aminocaproic acid.

Transfers from agarose gels.

Use the discontinuous buffer system *without* methanol.

2 Technical procedure

These explanations refer to the discontinuous buffer system. If a continuous buffer system is used, the anode solutions I and II and the cathode buffer are identical.

To avoid contaminating the buffer, blotting membranes and filter papers, rubber gloves should be worn to carry out the following operations.

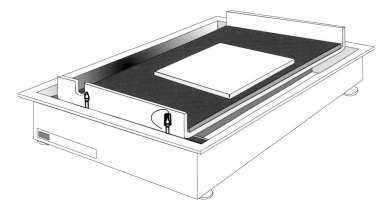

Fig. 3: Assembling the blotting sandwich on the graphite anode plate.

● Wet the graphite anode plate (with the red cable) with distilled water, remove the excess water with absorbent paper. *This enables a regular current to be obtained.*

● Cut the necessary filter papers (6 for the anode, 6 for the cathode, 6 per transfer unit) and the blotting membrane to the size of the gel. *or use a plastic frame as suggested before*

● Pour 200 mL of anode solution I into the staining tray;

● slowly impregnate 6 filter papers and place them on the graphite plate (anode) (Fig. 3); *avoid air bubbles*

● pour the anode solution I back into the bottle; *The buffer can be used several times.*

● pour 200 mL of the anode solution II into the staining tray;

● interrupt the electrophoresis or the isoelectric focusing; *see methods 7 and 8*

- wipe the kerosene off the bottom of the film;
- equilibrate the gels in the anode buffer I:

    SDS-PAGE: 5 min
    IEF: 2 min

SDS gels are soaked in metha-nol to prevent them from swel-ling.
It is easier to remove IEF gels from the blotting membrane af-ter blotting.

Gels on support films are left to "swim" on the top of the buffer with the gel surface on the bottom.

- Take the cooling plate from the electrophoresis chamber;

with the refrigeration tubing at-tached

- place the anode plate in the electrophoresis chamber, plug in the cables.

Tip!

The handling of a whole gel (25 × 10 cm) during removal is much easier if the gel is cut in two after equilibration, and both halves are put together again after pulling the FilmRemover wire through. A *single* blotting membrane can nevertheless be used for the whole gel.

Folds often form when a large support film is removed and it is difficult to get rid of them.

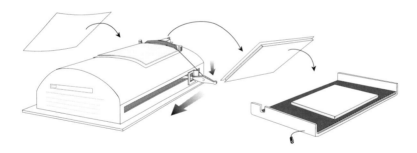

Fig. 4: Removing the support film and transferring the support film-gel-blotting membrane sandwich to the stack of anode paper.

- Place gels bound to support films with the film side down on the Film Remover; so that a short side is in contact with both the gel clamps;

The holding mechanism is now in the elevated position.

- Press on the lever;

Both teeth are now pressing on the edge of the film and hold it in place.

- Place the wire over the edge of the gel beside the clamps, hook it to the handle on the other side and push the lever down (Fig. 4);

The wire now has the mechani-cal tension necessary to separa-te the gel completely from the support film.

● grasp the handles with both hands and smoothly pull the wire towards you.

While doing this, press the lever down with a finger from the right hand so that it does not spring up.

Film bound gels:

● Briefly soak the blotting membrane in the anode buffer II and place it on the gel;

● slowly soak 3 filter papers in the buffer and place them on the prepared stack of filter papers;

● press on the top handle and lift the whole sandwich with the support film, turn in over and place it on the stack of filter papers;

For gels which are not bound to support film, proceed in the opposite way:

The equilibration step is usually not carried out for gels which are not bound to a support film.

● slowly soak 3 filter papers in buffer and place them on the prepared stack of filter papers;

● briefly soak the blotting membrane in the anode buffer II and place it on the stack of filter papers.

● Place the gel on the blotting membrane;

● pour off the anode buffer II;

This buffer should only be used once (kerosene) .

● rinse out the staining tray with distilled water and dry it with paper;

Reusable cathode buffer should not be contaminated with traces of anode buffer.

● pour 200 mL of cathode solution in the staining tray;

● pull the film away slowly, starting at one corner;

*see the **tip** above*

● soak 9 filter papers in the cathode buffer and place them on top.

When building a blot sandwich as described here, it is difficult to completely avoid air bubbles. They must therefore be pressed out with a roller (Fig. 5):

● Start in the middle and roll out in all four directions,

press in such a way that the buffer in the sandwich oozes out but is not completely pressed out. When the roller is removed, the buffer should be "drawn" back in;

If not enough pressure is app- lied, air bubbles will stay in the sandwich and no transfer will take place at these points. On the other hand if too much pres- sure is applied, the sandwich will be too dry and irregular.

● wet the graphite cathode (black cable) with distilled water; remove the excess water with absorbent paper;

see anode plate

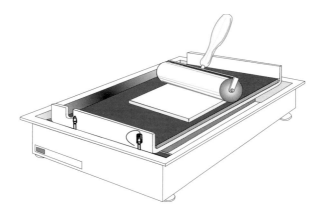

Fig. 5: Rolling out the air bubbles.

● place the cathode plate on the stack, plug in the cable;

● place the safety lid on the electrophoresis chamber and connect the cable to the power supply;

make sure that the polarity is correct

● blot at a constant current: 0.8 mA/cm^2

Higher currents result in a rise of the temperature of the gel and are not recommended.

Transfer conditions:

0.5 mm thick gel (250 cm^2) SDS pore gradient T = 8 to 20 %:				
Set:	200 mA const	max 10 V	max 5 W	20 °C
Read:	200 mA	3 V	1 W	

Some power supplies can be programmed to switch off after one mA, h-integration (133 mAh) for better reproducibility.

The blot does not warm up under these moderate conditions.

If thicker or more concentrated gels are blotted, the blotting time can be increased up to two hours; it is then recommended to press down the cathode plate with a 1 to 2 kg weight so that no electrolytic gas pockets form.

These gels have enough mechanical stability that they are not crushed.

Blotting native or focusing gels is quicker, because the proteins are in globular form and, for IEF, the gels have larger pores.

guideline: 30 min

● Switch off the power supply, unplug the cable;

● remove the safety lid and the cathode plate;

● take the blot sandwich apart;

● stain the gel with Coomassie Blue to check.

see methods 7 and 8.

● Before visualization, either dry the gel overnight or for 3 to 4 h at 60 °C in a heating cabinet.

The proteins bind more firmly to the film during drying and are not washed out during staining and specific detection.

Caution: *This treatment should not be used before detection of biological activity (zymogram techniques) since most enzymes lose their activity.*

3 Staining of blotting membranes

Amido Black staining:

Dissolve 10.1 g of Amido Black in 100 mL of methanol-glacial acetic acid-water (40:10:50 v/v);

● stain for 3 to 4 min in 0.1% Amido Black solution;

● destain in methanol-glacial acetic acid-water (25:10:65 v/v);

● air dry the nitrocellulose.

Reversible staining:

If a specific immuno- or glycoprotein identification test is to be carried out after the general detection, it is recommended that staining with Ponceau S (Salinovich and Montelaro, 1986) or Fast Green FCF is used .

Fast Green staining

● dissolve 0.1% (w/v) of Fast Green in 1% acetic acid;

Mild staining for proteins!

● stain for 5 min;

● destain the background with distilled water for 5 min;

complete destaining of the bands is achieved by incubating the film for 5 min in 0.2 mol/L NaOH.

This only works for nitrocellulose, in addition alkaline treatment increases antigen reactivity. See page 60: Sutherland and Skerritt (1986)

Blocking and immunological or lectin detection can now be carried out.

Indian Ink staining, Hancock and Tsang (1983):

Unfortunately Indian Ink is no longer produced. However, the description of this fine method is not removed from this issue of the book for the case it would again be produced, or an ink with comparable performance would be found.

For the sensitivity see Fig. 6 as a comparison with a gel stained with Coomassie Blue.It comes close to the sensitivity of silver staining of gels.

● *Soaking:* 5 min in 0.2 mol/L NaOH;

● *Washing:* 4 × 10 min with PBS-Tween (250 mL per wash, agitate).

PBS-Tween: 48.8 g of NaCl + 14.5 g of Na_2PO_4 + 1.17 g of NaH_2PO_4 + 2.5 mL of Tween 20, fill up to 5 L with distilled water.

● *Staining:* 2 h or overnight with 250 mL of PBS Tween + 2.5 mL of acetic acid + 250 mL of fountain pen ink ("Fount India");

Rocking platform! Filter the staining solution first.

● *Washing:* 2 × 2 min with water;

● *Drying:* air-dry.

Plastic embedding of indian ink stained blots does not work.

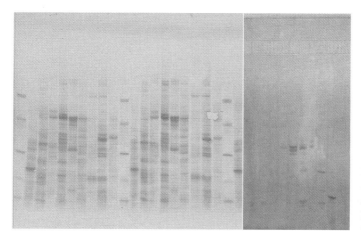

Fig. 6: *Left:* Blot of an SDS electrophoretic separation of legume seed extracts on nitrocellulose; Indian ink stain. *Right:* Coomassie stained gel containing the same protein pattern in the same concentration.

Plastic embedding of the blotting membrane

Nitrocellulose can be made totally and permanently transparent after staining or specific immunological detection methods, so that the result looks like a stained electrophoresis gel.

Pharmacia LKB Development Technique File No 230, PhastSystem® (1989).

The membrane is soaked in a monomer solution with the same refractive index as polymerized nitrocellulose. For permanent transparency a photo-initiator, which starts polymerizing in the presence of a UV light source, is added to the monomer solution.

The monomer chosen for this purpose is odorless and relatively safe.

Monomer solution: dissolve 0.5 g of benzoin methyl ether in 25 mL of TMPTMA. Store at room temperature in a brown flask or in the dark. The solution can be kept for about 2 weeks. For longer storage, keep it in the refrigerator at 4 °C to 8 °C.

Complete dissolution may take several hours. Avoid skin contact: TMPTMA is an irritant for the skin and mucous membranes.

● Dry the nitrocellulose completely;

● cut 2 sheets of PVC to a size larger than that of the blotting membrane;

sheets for the overhead projector

● pipette 0.3 mL of monomer solution on one of the plastic sheets;

or more depending on the size of the blotting membrane

● spread the solution so that it covers an area equal to the size of the membrane;

● stand the blotting membrane on its edge and slowly lower it on to the monomer solution so that the membrane is completely impregnated;

● apply a few drops of solution on to the membrane and place the second plastic sheet on top;

● carefully press out the air bubbles with a roller;

● irradiate both sides for about 15 s with a UV lamp;

● cut the sandwich to the size desired.

Method 9: IEF in immobilized pH gradients

The principles of isoelectric focusing in immobilized pH gradients are described in section I. This technique possesses a number of characteristics which make it clearly different from IEF in carrier ampholyte gradients:

see pages 52 and following as well as "Strategies for IPG focusing" at the end of this chapter

Advantages:

● The separations are more reproducible. The pH gradient is absolutely linear and remains so, it is not influenced either by salt or buffer ions or by the proteins of the sample.

There are no wavy lines, only absolutely straight iso-pH lines.

● The pH gradient is absolutely stable in time, it cannot drift.

Field strength and time are not limited.

● Basic proteins are better focused in immobilized pH gradients because the gradient remains stable.

Most problems of carrier ampholyte IEF occur in this range.

● Immobilized pH gradients are more reproducible, because they are produced by at most six different substances which are chemically well defined.

in contrast to several 100 carrier ampholyte homologues

● Since immobilized pH gradients can be custom-made and with very narrow pH ranges, very high resolution can be achieved.

$< \Delta pI = 0.001\ pH$

● Since only tertiary amines are used as buffering groups and no carrier ampholytes are present, low molecular weight proteins can be detected directly in the gel with ninhydrin or dansyl chloride.

Carrier ampholytes are also stained by these substances.

● Immobilized pH gradients are a part of the gel; therefore there are no edge effects and the separations can be carried out in narrow, individual gel strips.

This is especially practical for the first dimension of two-dimensional electrophoresis.

Disadvantages:

● The gel casting technique for IPG gels is more complicated.

pipetting the Immobiline, pouring the gradients

● Errors can occur during the preparation of the gels.

e.g. errors during pipetting.

● IPGs can only be made with polyacrylamide.

no agarose gels

● The separation takes more time.

because of the lower conductivity

● High voltages are necessary.

also because of the lower conductivity

● Many proteins do not enter the gel easily.

Tricks are sometimes necessary here.

1 Sample preparation

The best results are obtained when concentrated samples are diluted 1:3 or more before application. The loading capacity of Immobiline gels is higher than that of Ampholine gels. The limit is set by the concentration of the sample when the proteins penetrate the gel. When a large amount of proteins is to be applied, it is recommended to dilute the sample and apply it in several steps. In many cases it has proved useful to mix the sample with about 2% of Triton X-100 or polyethylenglycol.

For membrane proteins and proteins from lyophilized platelets, a significant increase in the yield of protein extraction by the addition of nondetergent sulfobetaines has been reported by Vuillard *et al.* (1995)

Vuillard L, Marret N, Rabilloud T. Electrophoresis. 16 (1995) 295-297.

When preparing cell lysates 8 mmol/L of PMSF (at 5×10^8 cell equivalents per mL of lysate solution) should be added as protease inhibitor (Strahler *et al.* 1987).

Strahler JR, Hanash SM, Somerlot L, Weser J, Postel W, Görg A. Electrophoresis. 8 (1987) 165-173.

Since Immobiline gels are extensively washed during their preparation, there are no more toxic monomers and catalysts.

2 Stock solutions

Immobiline® II 0.2 molar stock solutions:

pK 3.6 (a) 10 mL
pK 4.6 (a) 10 mL
pK 6.2 (b) 10 mL
pK 7.0 (b) 10 mL
pK 8.5 (b) 10 mL
pK 9.3 (b) 10 mL

(a) = acid,
(b) = basic

The solutions are stabilized against autopolymerization and hydrolysis and stay stable for at least 12 months when stored in the refrigerator (4 to 8 °C).

Immobilines® II should *not* be frozen!

acid Immobiline® II (a) dissolved in double-distilled water and stabilized with 5 ppm of polymerization inhibitor (hydrochinone monoethylether), basic Immobiline® II (b) dissolve in n-propanol

Acrylamide, Bis (T = 30%, C = 3%):

29.1 g of acrylamide + 0.9 g of Bis, make up to 100 mL with double-distilled water, or
reconstitute PrePAG Mix (29.1:0.9) with 100 mL of double-distilled water.

Dispose of the remains ecologically: polymerize with an excess of APS.

Beware! *Acrylamide and Bis are toxic in the monomeric form. Avoid skin contact, do not pipette by mouth. The solution stays stable for a week when stored in the dark at 4 °C (refrigerator). It can still be used for several weeks for SDS gels.*

The quality of the solutions is not as important for SDS gels as for IEF gels.

Ammonium persulfate solution (APS) 40% (w/v):
dissolve 400 mg of APS in 1 mL of double-distilled water.

4 mol/L HCl: 33.0 mL of HCl, make up to 100 mL with double-distilled water.

2 mmol/L acetic acid: 11.8 µL of acetic acid, make up to 100 mL with double-distilled water.

2 mmol/L Tris: 24.2 mg of Tris, make up to 100 mL with double-distilled water.

it remains stable for one week when kept in the refrigerator (4 °C)

3 Immobiline recipes

To cast the pH gradient two solutions are prepared, an acid, dense (with glycerol) one and a basic, light one. The acid solution is the acid end point of the gel and the basic one, the basic end point. The standard gel thickness is 0.5 mm; 7.5 mL of each starting solution are necessary for a gel. The recipes in table 1 and 2 are calculated for a optimum ionic strength of about 5 mmol/L, the pH and pK values are given at 10 °C.

From the beginning, it was decided to choose the acid solution as the denser one for Immobiline gels. This means that the acid end of the gel is always at the bottom.

Custom-made pH gradients

If custom-made pH gradients are needed to optimize the resolution (e.g. for preparative uses) the Immobiline volumes of both starting solutions can be calculated with a computer, using the software by Altland (1990) or by Giaffreda *et al.* (1993).
The recipes can also be deduced graphically (Fig. 1):

● Choose the gradient which corresponds best to the *desired* gradient from Tab. 1 or Tab. 2.

For example:
desired: pH 4.3 to 6.2,
chosen: pH 4.0 to 7.0 (Tab. 2)

● Report the volumes of the Immobiline acid and basic starting solutions (µL) on a scale of the *chosen* gradient on graph paper.

● Connect the quantities (µL) of the related Immobilines with lines.

Linear gradient!

● Mark the acid and basic end points of the *desired* gradient on the pH gradient scale and draw vertical lines from this point.

● Read the corresponding Immobiline volumes at the intersection of the vertical lines and the joining lines. This is the recipe for the starting solutions of the *desired* gradient (Fig. 1).

Sometimes several recipes are available for one gradient. Various formulations are then possible for a gradient, yet the resulting gradients will be identical.

Tab 1: Narrow pH gradients: Immobiline quantities for 15 mL of the starting solutions (2 gels) 0.2 mol/L Immobiline in µL.

Acidic, dense solutions (D.S.) Immobiline pK						pH gradient	Basic, light solution (L.S.) Immobiline pK					
3.6	4.6	6.2	7.0	8.5	9.3		3.6	4.6	6.2	7.0	8.5	9.3
—	904	—	—	—	129	3.8— 4.8	—	686	—	—	—	477
—	817	—	—	—	141	3.9— 4.9	—	707	—	—	—	525
—	755	—	—	—	157	4.0— 5.0	—	745	—	—	—	584
—	713	—	—	—	177	4.1— 5.1	—	803	—	—	—	659
—	689	—	—	—	203	4.2— 5.2	—	884	—	—	—	753
—	682	—	—	—	235	4.3— 5.3	—	992	—	—	—	871
—	691	—	—	—	275	4.4— 5.4	—	1133	—	—	—	1021
—	716	—	—	—	325	4.5— 5.5	—	1314	—	—	—	1208
562	600	863	—	—	—	4.6— 5.6	—	863	863	—	—	105
458	675	863	—	—	—	4.7— 5.7	—	863	863	—	—	150
352	750	863	—	—	—	4.8— 5.8	—	863	863	—	—	202
218	863	863	—	—	—	4.9— 5.9	—	863	863	—	—	248
158	863	863	—	—	—	5.0— 6.0	—	863	803	—	—	338
113	863	863	—	—	—	5.1— 6.1	—	863	713	—	—	443
1251	—	1355	—	—	—	5.2— 6.2	337	—	724	—	—	—
1055	—	1165	—	—	—	5.3— 6.3	284	—	694	—	—	—
899	—	1017	—	—	—	5.4— 6.4	242	—	682	—	—	—
775	—	903	—	—	—	5.5— 6.5	209	—	686	—	—	—
676	—	817	—	—	—	5.6— 6.6	182	—	707	—	—	—
598	—	755	—	—	—	5.7— 6.7	161	—	745	—	—	—
536	—	713	—	—	—	5.8— 6.8	144	—	803	—	—	—
486	—	689	—	—	—	5.9— 6.9	131	—	884	—	—	—
447	—	682	—	—	—	6.0— 7.0	120	—	992	—	—	—
416	—	691	—	—	—	6.1— 7.1	112	—	1133	—	—	—
972	—	—	1086	—	—	6.2— 7.2	262	—	—	686	—	—
833	—	—	956	—	—	6.3— 7.3	224	—	—	682	—	—
722	—	—	857	—	—	6.4— 7.4	195	—	—	694	—	—
635	—	—	783	—	—	6.5— 7.5	171	—	—	724	—	—
565	—	—	732	—	—	6.6— 7.6	152	—	—	771	—	—
509	—	—	699	—	—	6.7— 7.7	137	—	—	840	—	—
465	—	—	683	—	—	6.8— 7.8	125	—	—	934	—	—
430	—	—	684	—	—	6.9— 7.9	116	—	—	1058	—	—
403	—	—	701	—	—	7.0— 8.0	108	—	—	1217	—	—
381	—	—	736	—	—	7.1— 8.1	103	—	—	1422	—	—
1028	—	—	750	750	—	7.2— 8.2	548	—	—	750	750	—
983	—	—	750	750	—	7.3— 8.3	503	—	—	750	750	—
938	—	—	750	750	—	7.4— 8.4	458	—	—	750	750	—
1230	—	—	—	1334	—	7.5— 8.5	331	—	—	—	720	—
1037	—	—	—	1149	—	7.6— 8.6	279	—	—	—	692	—
885	—	—	—	1004	—	7.7— 8.7	238	—	—	—	682	—
764	—	—	—	893	—	7.8— 8.8	206	—	—	—	687	—
667	—	—	—	810	—	7.9— 8.9	180	—	—	—	710	—
591	—	—	—	750	—	8.0— 9.0	159	—	—	—	750	—
530	—	—	—	710	—	8.1— 9.1	143	—	—	—	810	—
482	—	—	—	687	—	8.2— 9.2	130	—	—	—	893	—
443	—	—	—	682	—	8.3— 9.3	119	—	—	—	1004	—
413	—	—	—	692	—	8.4— 9.4	111	—	—	—	1149	—
389	—	—	—	720	—	8.5— 9.5	105	—	—	—	1334	—
1208	—	—	—	—	1314	8.6— 9.6	325	—	—	—	—	716
1021	—	—	—	—	1133	8.7— 9.7	275	—	—	—	—	691
871	—	—	—	—	992	8.8— 9.8	235	—	—	—	—	682
753	—	—	—	—	884	8.9— 9.9	203	—	—	—	—	689
659	—	—	—	—	803	9.0—10.0	177	—	—	—	—	713
584	—	—	—	—	745	9.1—10.1	157	—	—	—	—	755
525	—	—	—	—	707	9.2—10.2	141	—	—	—	—	817
478	—	—	—	—	686	9.3—10.3	129	—	—	—	—	903
440	—	—	—	—	682	9.4—10.4	119	—	—	—	—	1017
410	—	—	—	—	694	9.5—10.5	111	—	—	—	—	1165

Tab. 2: Wide pH gradients: Immobiline quantities for 15 mL of the starting solutions (2 gels), 0.2 mol/L Immobiline in µL.

Acidic, dense solution (D.S.) Immobiline pK						pH gradient	Basic, light solution (L.S.) Immobiline pK					
3.6	4.6	6.2	7.0	8.5	9.3		3.6	4.6	6.2	7.0	8.5	9.3
299	223	157	—	—	—	3.5– 5.0	212	310	465	—	—	—
569	99	439	—	—	—	4.0– 6.0	390	521	276	—	—	722
415	240	499	—	—	—	4.5– 6.5	—	570	244	235	—	297
69	428	414	—	—	—	5.0– 7.0	—	474	270	219	—	320
—	450	354	113	—	—	5.5– 7.5	347	—	236	287	284	—
435	—	323	208	44	—	6.0– 8.0	286	—	174	325	329	—
771	—	276	185	538	—	6.5– 8.5	192	—	153	278	362	—
1349	—	—	272	372	845	7.0– 9.0 ·	484	—	—	232	189	546
668	—	—	445	226	348	7.5– 9.5	207	—	—	925	139	346
399	—	—	364	355	94	8.0–10.0	91	—	—	329	366	289
578	110	450	—	—	—	4.0– 7.0	302	738	151	269	—	876
702	254	416	133	346	—	5.0– 8.0	175	123	131	345	346	—
779	—	402	93	364	80	6.0– 9.0	241	—	161	449	237	225
542	—	—	378	351	—	7.0–10.0	90	—	—	324	350	280
588	254	235	117	170	—	4.0– 8.0	—	554	360	142	334	288
830	582	218	138	795	122	5.0– 9.0	—	249	263	212	292	230
941	—	273	243	260	282	6.0–10.0	100	—	333	361	239	326
829	235	232	22	250	221	4.0– 9.0	147	424	360	296	71	663
563	463	298	273	227	127	5.0–10.0	21	59	34	420	310	273
1102	—	455	89	334	—	4.0–10.0	—	114	50	488	157	357

The graphic interpolation is sometimes also used for empirical pH gradient optimization:

A sample is separated in a chosen Immobiline pH gradient. If a better resolution of a definite group of bands within the gradient is desired, draw the pH scale of the gradient used to the length

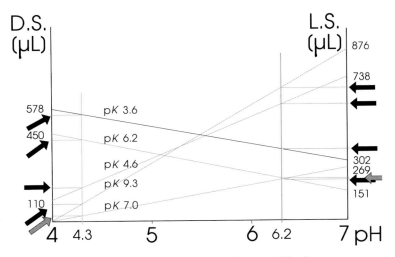

Fig. 1: Graphic interpolation of a custom-made immobilized pH gradient pH 4.3 – 6.2. Ordinate: Immobiline volumes.

of the separation distance on graph paper. Place the stained gel on the drawing so that the trace and the scale are superimposed (do not forget the orientation of the gel: acid end - anode and basic end - cathode!). Draw the end points of the new gradient desired (to the top and the bottom of the group of bands) and continue as described above.

4 Preparing the casting cassette

Before use, treat the inner side of the spacer with Repel Silane so that the gel can be removed after separation.

The "spacer" is the glass plate with the 0.5 mm thick U-shaped silicone rubber gasket.

Under the fume hood drop a few mL of Repel Silane on the surface of the plate and spread it over the whole surface with tissue (wear rubber gloves). When the solution has dried, wash the plate under running water and then rinse it with distilled water (to remove the chloride ions which result from the evaporation of the solvent).

Making the surface hydrophobic.

This treatment only needs to be carried out once, test with a drop of water.

The gel is covalently polymerized on a GelBond PAG film. Place a clean glass plate on absorbent paper and wet it with a few mL of water. Place a GelBond PAG film with the untreated, hydrophobic side down on the glass plate and press it on with a roller (Fig. 2).

*It is a good idea to mark the **hydrophobic** side of the film with a black marker: gradient pH from ... to ... acid (+) and basic (-) side.*

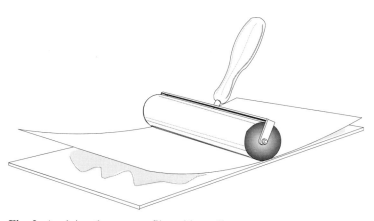

Fig. 2: Applying the support film with a roller.

A thin layer of water then forms between the glass plate and the film and holds them together by adhesion. The excess water which runs out is absorbed by the tissue.

To facilitate pouring the gel solution, let the film overlap the length of the glass plate by about 1 mm.

Place the spacer over the glass plate and clamp the cassette together (Fig. 3). Cool the casting cassette in the refrigerator at 4 °C for about 10 min before filling it, this delays the onset of polymerization. This last step is essential since it takes 5 to 10 min for the poured gradient in the 0.5 mm thick layers to settle horizontally.

In a warm laboratory in the summer, also cool down the starting solution to 4 °C.

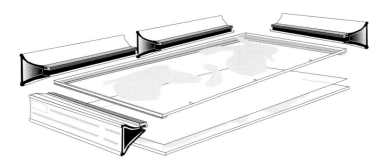

Fig. 3: Assembling the gel cassette.

5 Preparing the pH gradient gels

As can be seen from the recipes, both starting solutions are brought to an identical optimum pH value of 7 with HCl or TEMED. This is especially necessary for broad and relatively high or low pH ranges. The effectiveness of the polymerization depends on the pH; this means that when pouring a pH gradient, different or extreme pH values will result in varying rates of polymerization of the Immobiline within the gel layer. The pH gradients occurring in the polymerized gels would then not correspond to the ones calculated.

NaOH can also be used instead of TEMED, but the use of TE-MED is easier and not harmful. The substances used for the titration are not polymerized in the gel and can be removed by washing.

The quantities of HCl and TEMED listed in Tab. 3 are only valid for the recipes in this example. For other gradients, pH 7 should be cautiously approached in 5 µL-steps.

Tab. 3: Recipes for starting solutions for two 0.5 mm thick IPG gels pH 4 to 10, $T = 4\%$, $C = 3\%$.

	Acidic solution pH 4.0 V (µL)	Basic solution pH 10.0 V (µL)
Glycerol (87 %)	4 300	800
Immobiline pK 3.6	1 102	-
Immobiline pK 4.6	-	114
Immobiline pK 6.2	455	50
Immobiline pK 7.0	89	488
Immobiline pK 8.5	334	157
Immobiline pK 9.3	-	357
Acrylamide, Bis solution	2 000	2 000
make up with distilled water	→ 15 000	→ 15 000
TEMED (100 %)	7,5	7,5
mix carefully		
and measure the pH		
4 mol/L HCl	-	20
TEMED (100 %)	15 mL *)	
	(experimental values)	

pH 4 to 10 is the example given, for other gels the corresponding Immobiline quantities are to be taken from Tab. 1 or Tab.2. T=4% and C=3% are ideal for most gels.

The precision of pH paper is sufficient here.
titrate to pH 7

*) can already be taken into consideration when TEMED is added above:
$\Sigma V = 22.5$ µL

Only add APS in the mixing chamber so the gel has more time to settle horizontally.

Pouring the gradient

a) Setting up the casting apparatus

A *gradient mixer* is used to prepare the gradient. It consists of two communicating chambers (Fig. 4).The dense, acid solution and a stirrer bar are placed in the front cylinder, the *mixing chamber*. The rear cylinder, the *reservoir*, contains the light basic solution. 25 % glycerol is added to the dense solution and 5 % to the light one so that it is easier to overlay the gel solution in the cassette with water before polymerization.

It is not recommended to use sucrose for the density gradient of thin gels because the solution would be too viscous.

The gel solutions flow to the cassette through the tubing under the influence of gravity, no pump is necessary.

Position the bottom of the mixing chamber 5 cm above the top of the cassette with a laboratory platform to reach the

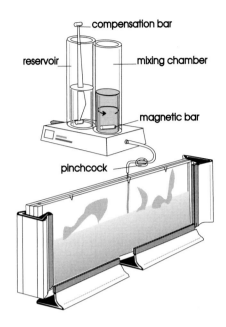

For reproducible gradients the outlet of the gradient mixer must always be on the same level above the edge of the gel cassette.

Fig. 4: Pouring the linear pH gradient.

optimum flow rate (cut the tubing to the correct length).

Because of the differences in the density of the solutions, the dense solution would flow back into the gradient mixer when the connecting tubing is opened, this can be prevented with the *compensating bar*.

It also compensates the volume of the stirring bar and helps to mix APS into the light solution.

Add the light and the dense solutions directly into the gradient mixer in the following order:

● pour 7.5 mL of light solution into the reservoir;

● briefly open the valve to fill the connecting channel;

● pour 7.5 mL of dense solution into the mixing chamber.

When everything is ready, remove the casting cassette from the refrigerator and place it beside the gradient mixer. The glass plate with the film should be facing you.

● Add APS (40%) to the starting solutions:

to the dense solution (7.5 mL) 7.5 μL

to the light solution (7.5 mL) 7.5 μL

b) Filling the cassette

● Add APS to the starting solutions:

mix it well by vigorously stirring with the stirring bar in the mixing chamber and the compensating stick in the reservoir;

The linear gradient is established during pouring by continuous overlayering of the gel solution which always becomes less dense.

● insert the tubing in the middle notch of the glass plate;

● set the stirrer at a moderate speed;

● open the valve connecting the chambers;

● open the pinchcock.

After a few minutes, when the liquid level has reached a height of about 1 cm below the edge of the cassette, the gradient mixer should be empty.

Immediately rinse the gradient mixer with double-distilled water.

● **Important:** Overlay the gel with about 0.5 mL of double-distilled water, so that the edge of the basic part becomes even and to prevent oxygen from diffusing into the gel;

Water works best; nothing else, e.g. butanol, has proved comparable.

● let the gel polymerize for 10 min at room temperature, so the density gradient can settle;

● polymerize the gel for 1 h in a heating cabinet at 37 °C;

Make sure the cassette is horizontal.

● cool the cassette under running water and remove the gel from the cassette.

c) Washing the gel

Immobiline gels exhibit low conductivity because the buffering groups of the pH gradient are firmly bound in the gel and cannot move freely like carrier ampholytes. Ions, such as TEMED and APS, which are not co-polymerized cannot be transported electrophoretically out of the gel during IEF. Therefore Immobiline gels must be washed with double-distilled water after polymerization.

The gels swell during washing. Sometimes in a specific pH interval they acquire a characteristic surface structure (snake skin). But the surface becomes smooth again once the gel is dry.

The gels are then dried and rehydrated in a cassette before use. If the gels were used directly after drying, the excess water would seep out of the gel during IEF ("sweating"). The gels should be completely dried, since reswelling and drying result in an irregular gel thickness. They are rehydrated in a cassette (Fig. 6).

Using a cassette has the additional advantage that one has a quantitative control over the additive concentration.

● Wash the gel in double-distilled water:

agitate 4 times for 15 min in 300 mL of double-distilled water each time.

If a MultiWash appliance is used, deionize by pumping double-distilled water through the mixed bed ion-exchange cartridge for about 60 min (see Fig. 5).

● Equilibrate the gel for 15 min in 1.5% (v/v) glycerol;

so that it does not roll up during drying

● dry the gel at room temperature with a fan or a ventilator;

in a dust free cabinet if possible

● as soon as possible, when the gel is dry, cover it with Mylar film and store in the freezer in plastic bags (Rossmann and Altland, 1987).

Rossmann U, Altland K. Electrophoresis. 8 (1987) 584-585.

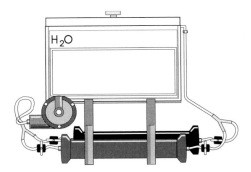

Fig. 5: MultiWash. Gel staining and washing appliance.

d) Storage

The dry gels should be stored at −20 °C, so that their swelling properties are retained.

e) Rehydration

The dry gels are reconstituted in a reswelling cassette (Loessner and Scherer, 1992) or in a rehydration tray in the calculated volume of rehydration solution (Fig. 6), so that the gel reswells evenly and to ensure a quantitative control over the additive concentration.

Loessner MJ, Scherer S. Electrophoresis.13 (1992) 461-463.

A few examples for the rehydration of Immobiline gels are shown in Tab. 4.

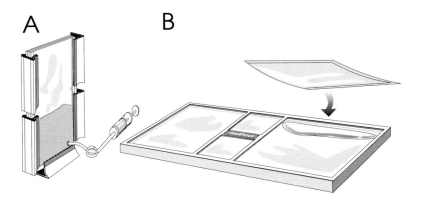

Fig. 6: Rehydration of an IPG gel in a vertical cassette (A) or alternatively in the GelPool (B).

Tab. 4: Rehydration of Immobiline gels.

Rehydration solution	Time h	Samples	
Dist. water	1	Water soluble proteins, enzymes	
2 mmol/L acetic acid (anodal sample appl.: - prevents lateral band spreading)	1	Water soluble proteins, enzymes	
2 mmol/L Tris (cathod. sample appl.: - prevents lateral band spreading)	1	Water soluble proteins, Enzyme	*Bjellqvist B, Linderholm M, Östergren K, Strahler JR. Electrophoresis 9 (1988) 453-462.*
25 % (v/v) glycerol	1	Serum for determin. of PI, GC, TF	*LKB Application Note 345 (1984).*
8 mol/L urea	2	Poorly soluble proteins , protein complexes,, Plasma: prealbumin	*Altland K, Banzhoff A, Hackler R, Rossmann U. Electrophoresis. 5 (1984) 379-381.*
8 mol/L urea, 0,5 % (g/v) Ampholine, 2 % (v/v) 2-Mercaptoethanol or 50 mmol/L DTT	over-night	Dried blood, plasma, erythrocyte lysates, globins,	*Rossmann and Altland (1987).*
8 mol/L urea, 0.5 % (w/v) Ampholine, 30 % (v/v) glycerol	over-night	Serum VLDL for the determin. of apolipoprotein E	*Baumstark M, Berg A, Halle M, Keul J. Electrophoresis. 9 (1988) 576-579.*
5 mol/L urea, 20 % (v/v) glycerol	2 h	alcohol soluble proteins	*Günther S., Postel, W., Weser, J, Görg A. In: Dunn MJ, Ed. Electrophoresis'86. VCH Weinheim (1986) 485-488.*
0.5 % (w/v) Ampholine	1	PGM, time sensitive enzymes,	*LKB Appl.Notes 345 and 373 (1984).*
4 % (w/v) Ampholine, 2 % (v/v) Nonidet NP-40	over-night	Membrane proteins	*Rimpilainen M, Righetti PG. Electrophoresis. 6 (1985) 419-422.*
4 % (w/v) Ampholine,	over-night	Alkaline Phosphatase	*Sinha PK, Bianchi-Bosisio A, Meyer-Sabellek W, Righetti PG. Clin Chem. 32 (1986) 1264-1268.*
up to 9 mol/L urea, 0.5 % (v/v) Nonidet NP-40	over-night	Hydrophobic proteins complex protein mixtures, 2D-electrophoresis	*Görg A, Postel W, Weser J, Günther S, Strahler JR, Hanash SM, Somerlot L. Electrophoresis. 8 (1987a) 45-51.*
8 mol/L urea, 2 % (v/v) Nonidet NP-40, 10 mmol/L DTE	16-18	2D electrophoresis of myeloblast lysates	*Hanash SM, Strahler JR, Somerlot L, Postel W, Görg A. Electrophoresis. 8 (1987) 229-234.*

If merely a few samples are to be separated, only part of the IPG gel needs to be rehydrated. The unused part should be stored at −20 °C tightly packed.

For 2D electrophoresis the gel is cut into individual strips with a paper cutter before rehydration (Hanash *et al.* 1987).

see method 10

To optimize additive concentration or to investigate conformational changes and ligand binding properties of certain proteins with regards to additive concentrations, an *additive gradient* perpendicular to the immobilized pH gradient can be established (Altland *et al.* 1984).

The rehydration cassette can also be used to establish immobilized pH gradients for long separation distances (up to 25 cm).

But it is not wise to have an Ampholine gradient perpendicular to the pH gradient, since a levelling of the concentration gradient occurs in the gel during IEF because of the high conductivity of the carrier ampholytes.

To prevent the surface of the gel from sticking to the glass (with the U-shaped gasket), the glass plate should first be coated with Repel Silane.

After rehydration, the gel surface should be dried, that is the whole solution should be soaked up by the gel. If the reswelling solution contains a high concentration of a non-ionic detergent (e.g. 1% Triton X-100), the surface may be slightly sticky after rehydration, in which case, the surface should be dried with precision wipe.

It is important to *orient* Immobiline gels properly - the acidic end at the anode (Fig. 8).

The *basic end* of the gel is easy to recognize by:
1) the wavy edge of the gel (the anodic edge is straight);
2) the wider film edge (ca 1.5 cm).

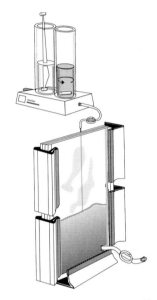

Fig. 7: Rehydrating in an additive gradient.

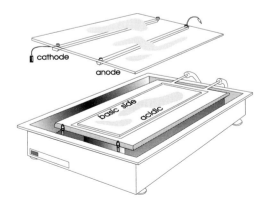

Fig. 8: Placing the Immobiline gel on the cooling plate of the separation chamber.

Use kerosene or DC-200 silicone oil between the cooling plate and the support film. If other fluids are used (e.g. distilled water, 1% Triton X-100) discharges or sparks can occur at the edge of the support film, since very high voltages are used with Immobiline gels.

6 Isoelectric focusing

Sample application

No prefocusing is done with Immobiline gels, the samples can be directly applied in the form of drops (10 to 20 µL on the surface). It has proved best if the gel/sample contact surface is as small as possible especially if the gel and/or sample contain non-ionic detergents (Strahler *et al*. 1987).

For separating hydrophobic proteins in the presence of 8 mol/L urea and 0.5 % NP-40 application in preformed sample wells or into holes, which have been punched into the gel (penetrating the support film) have proved to be suitable (Loessner and Scherer, 1992).

This has reduced lateral band spreading significantly.

Large sample volumes can also be applied with cut-out 2 mm thick silicone rubber frames (Hanash *et al*. 1987), cut-out sample application strips or silicone tubing, which are placed on the gel (Fig. 9). For IPG as well it is recommended to carry out a step test to find the optimum sample application point (s. method 6).

The low conductivity of IPGs should be taken into consideration (see the end of this chapter).

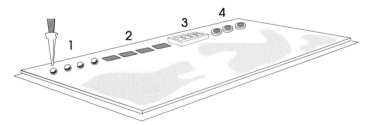

Fig. 9: Possibilities for sample application for IPG. - (1) droplets (2) pieces of cellulose (3) sample application mask (4) cut-off silicone tubing.

When many samples are focused together, silicone rubber application masks with funnel shaped holes and grooves on the bottom are particularly efficient (Pflug and Laczko, 1987).

Pflug W, Laczko B. Electrophoresis. 8 (1987) 247-248.

Electrode solutions

For immobilized pH gradients both electrode strips only have to be soaked in distilled water. This saves work, rules out mistakes, increases the absorption capacity of the strips for salt ions and reduces the field strength at the beginning (better sample entry!).

Focusing conditions

Normally focusing is carried out at 10 °C, lower (sensitive enzymes) or higher (e.g. 8 mol/L urea) temperatures being used for special applications. The pH values indicated on the packages are only valid at 10 °C. Many additives, e.g. urea, shift the pH value.

See also: "strategies for IPG focusing" at the end of this chapter.

The *separation times* depend on the steepness of the pH gradient, the additives, molecular sizes and power supply settings.

The *current* is set so that the field strength in the gel is low at first (40 V/cm); if the field strength is too high at the beginning some of the proteins will precipitate.

Make sure that the anodic cable is plugged in in front.

Example:

Whole IPG gel, pH 4.0 to 7.0, reswollen in double-distilled water, 10 °C:

Maximal value for the power supply: 3000 V, 1.0 mA, 5.0 W

Focusing time: 5 h

● After IEF, switch off the power supply and open the safety lid.

● The proteins can now be stained or blotted. Should problems occur, consult the trouble-shooting guide in the appendix.

Consult the strategy graph at the end of the instructions.

Measuring the pH gradient:

The pH values cannot be measured with a pH meter because of the low conductivity. However, the pH gradient can easily be determined with marker proteins.

7 Coomassie and silver staining

Colloidal Coomassie staining

The results are relatively quickly visible with this method. Few steps are necessary, the staining solution is stable, there is no background staining, oligopeptides (10 to 15 amino acids) which are not fixed by other methods can be detected. In addition, the solution is almost odorless (Diezel et al. 1972; Blakesley and Boezi,1977).

This method is especially recommended for Immobiline gels, because all other Coomassie Blue stains result in strongly colored backgrounds.

Preparation of the staining solution: dissolve 2 g of Coomassie G-250 in 1 L of distilled water and add 1 L of sulfuric acid (1 mol/L or 55.5 mL of conc. H_2SO_4 per L). After stirring for three hours filter (folded filter), add 220 mL of sodium hydroxide (10 mol/L or 88 g in 220 mL). Finally add 310 mL of a 100% (w/v) TCA solution and mix well, the solution will turn green.

Fixing, staining: 3 h at 50 °C or overnight at room temperature in the colloidal solution;

Washing out the acid: soak in water for 1 or 2 h, the green bands will become more intense.

5 minute silver staining of dried gels

This method can be used to stain dried gels already stained with Coomassie (the background must be completely clear) to increase the sensitivity or else for direct staining of an unstained gel after pretreatment (Krause and Elbertzhagen, 1987). A particular advantage of this method lies in the fact that no proteins or peptides are lost. During other silver staining methods they often diffuse out of the gel because they cannot be irreversibly fixed.

Pretreatment of unstained gels:

- fix for 30 min in 20% TCA,

- wash for 2 × 5 min in 45% methanol, 10% acetic acid,

- wash for 4 × 2 min in distilled water,

- impregnate for 2 min in 0.75% glycerol,

- air dry.

Solution A: 25 g of Na_2CO_3, 500 mL of double distilled water;
Solution B: 1.0 g of NH_4NO_3, 1.0 g of $AgNO_3$, 5.0 g of tungsto-silisic acid, 7.0 mL of formaldehyde solution (37%), make up to 500 mL with double-distilled water.

Silver staining:

Mix 35 mL of solution A with 65 mL of solution B *before use*, immediately soak the gel in the resulting whitish solution and incubate while agitating until the desired intensity is reached. Briefly rinse with distilled water.

Stop with 0.05 mol/L glycine. Remove the remains of metallic silver from the back of the film and the gel with a cotton swab.

Air dry.

Densitometry

Proceed exactly as described in method 6 for the densitometric evaluation of the gels.

Blotting

Electroblotting of Immobiline gels has become much easier, now that the Film Remover exists to remove the GelBond PAG film (see method 8).

This method is mostly used for the sequencing of proteins, since the results cannot be falsified by carrier ampholytes blotted at the same time (Aebersold *et al.* 1988).

Practical tip

In the next paragraph (strategies for IPG focusing) it is recommended, in specific situations, to cut out strips of gel between the sample tracks. Fig. 10 shows how the sample tracks of an IPG gel can be separated easily and precisely. The rehydrated gel is placed on the line template (see appendix). Two glass plates with U-shaped gaskets are placed along the long edges of the support film and finally a simple glass plate is laid on top. Strips of gel are scraped out with a 4 or 5 mm thick spatula.

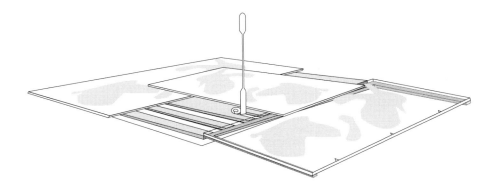

Fig. 10: Scraping out strips of gel between sample tracks of an IPG gel.

8 Strategies for IPG focusing

Sample application: concentration test

Burning line between sample application points

Cut out 5 mm wide gel strips between the tracks (Fig. 10) and apply the sample.

step trial test

The sample has a higher conduc tivity than the gel.

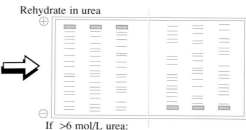

Lateral band broadening

that also helps

also a conductivity problem

Anodal   Cathodal
sample application

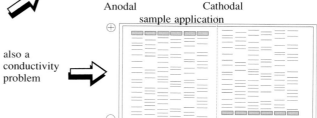

Rehydrate the gel in
2 mmol/L acetic acid   2 mmol/L Tris

The proteins smear on the surface

Part of the proteins tend to ag- gregate be- cause of the low conductivi- ty and do not enter the gel.

Rehydrate in urea

If >6 mol/L urea:
add 0.1 to 0.5% non-ionic detergent

Oxidation sensitive proteins e.g. hemoglobin

The gel is oxidized twice:

1. during polymerization (by APS)
2. during drying.

The proteins are oxidi- zed during the run.

1. First wash the gel for 30 min in 0.1 mol/L ascorbic acid, titrate with 1 mol/L Tris to pH 4.5 or in 2 mmol/L DTT solution and then in double-distilled water

or

2. Rehydrate the gel in 10 mmol/L DTT or in 2% 2- mercaptoethanol.

Method 10: High resolution 2D electrophoresis

The principle of high resolution 2D electrophoresis is descri- *see page 37 "Iso-Dalt" system*
bed in part 1. It has usually been carried out according to the *(Anderson and Anderson, 1984)*
method of O'Farrell (1975) or Klose (1975): first-dimensional
urea-IEF run in vertical cylindrical gels and in the second dimension, SDS-PAGE in vertical slab gels.

Main problems with the traditional first dimension were: *In addition basic carrier ampho-*
Poor reproducibility of the pattern; *lytes form complexes with SDS*
Loss of the basic proteins and some of the acidic ones. *which cover part of the pattern.*

IPG-Dalt
1st Dimension: By using immobilized pH gradients in indivi- *The immobilized pH gradient*
dual gel strips for the first dimension, stationary 2D electropho- *does not drift and is not influen-*
retic patterns are obtained over the whole pH spectrum (Görg *et* *ced by the sample composition.*
al. 1988a, Hanash and Strahler, 1989); no SDS-carrier ampholy-
te clouds occur. Because the focusing gel is polymerized on a *Blomberg A, Blomberg L, Nor-*
film, there is no change in length after the transfer. The repro- *beck J, Fey SJ, Larsen PM,*
Roepstorff P, Degand H, Boutry
ducibility is substantially increased, which has benn demonstra- *M, Posch A, Görg A. Electro-*
ted by an inter-laboratory comparison (Blomberg *et al.*1995). *phoresis. 16 (1995) 1935-1945.*

*2nd Dimension:*The second-dimensional run is much less *This chapter "Method 10"*
problematic. The results of the separation by SDS-PAGE do not *describes the procedure perfor-*
med in the horizontal plane.
depend on whether a vertical or horizontal technique is used. A *The instruction for the vertical*
pore gradient or a homogeneous separation gel with a correspon- *technique can be found in the*
ding acrylamide concentration are chosen depending on the *last chapter: "Method 15: Verti-*
composition of the sample. *cal PAGE".*
The following points should be considered when choosing the
optimal technique for the second dimension:

Vertical technique:

● For multiple sample runs, large buffer tanks for 10 or 20 gels *Anderson and Anderson (1984).*
placed in parallel are employed (IsoDalt system).

● The use of thick gels is necessary for semi-preparative *Eckerskorn and Lottspeich*
separations. *(1989).*

● A vertical chamber already available can be used.

Horizontal technique:

● The equilibrated IEF strip is placed on the gel without any
embedding with agarose .

● The size of the gel stays constant because of the support film
and it does not swell to a trapezoidal shape during staining.

● Gel casting and handling of the gels is easier in horizontal
systems.

The *horizontal technique* will be described here, because both dimensions can be run with one single piece of equipment.

The techniques of gel preparation for immobilized pH gradient and horizontal SDS pore gradient gels are described in methods 8 and 10 respectively. The following instruction contains optimized recipes for high resolution 2D electrophoresis.

Hint: Since immobilized pH gradient gels pick up charges and cause electroendosmotic effects in SDS gels (Westermeier *et al.* 1983), special precautions must be taken for the IPG-Dalt method when equilibrating the IEF strips and for the starting conditions of the second-dimensional run (Görg *et al.* 1985):

● restricted flow of hydration water,
● low field strength at the beginning,
● removal of the IPG strips after one hour.

Westermeier R, Postel W, Weser J, Görg A. J Biochem Biophys Methods. 8 (1983) 321-330.

Görg A, Postel W, Günther S, Weser J. Electrophoresis. 6 (1985) 599-604.

1 Sample preparation

For 2D electrophoresis, complex protein mixtures should be completely solubilized, also hydrophobic proteins. Complexes must be disaggregated, the polypeptide chains must be completely unfolded. The sample should have a very low salt content, no lipids and phenols should be present.

The anionic detergent SDS has been traditionally used to solubilize many proteins even hydrophobic ones. With samples prepared this way, IEF can only work − if it works at all, when the sample is applied to the cathode.

The negatively charged protein-SDS micelles must be broken up in presence of urea during the run.

When the samples are applied to the cathode, problems arise with 2-mercaptoethanol which is present in the solubilizing mixture. At high concentrations it overbuffers part of the gradient in the alkaline region.

It is therefore better to apply the sample at the anode since fewer proteins aggregate (Dunn and Burghes, 1983).

According to the present state of the art (Görg et al. 1995), the combination of urea, the zwitterionic detergent CHAPS, and a heterogeneous carrier ampholyte mixture has proved to be the best way to bring a complex protein mixture completely and reproducibly into solution.

Görg A, Boguth G, Obermaier C, Posch A, Weiss W. Electrophoresis. 16 (1995) in press.

Solubilizing mixture:

9 mol/L urea	5.4 g
2% (v/v) CHAPS	200 mg
2% (v/v) 2-mercaptoethanol	200 µL
0.8% (w/v) Pharmalyte pH 3 to 10	200 µL
8 mmol/L PMSF (protease inhibitor)	

make up to 10 mL with double-distilled water.

Prepare the solution freshly, shake to dissolve the urea, do not warm it.

For membrane proteins and proteins from lyophilized plate-lets, the addition of nondetergent sulfobetaines should be tried (Vuillard *et al.* 1995).

● Optimum protein concentration for sample application:

about 10 mg/mL of sample solution, this corresponds to 60-100 μg of protein/20 μL of applied sample (inclusive Dextran gel, see below) per IPG strip.

Silver staining concentration

In addition, to prevent the proteins from aggregating when they penetrate the gel, the sample is applied with a granular Dextran gel:

This decreases the mobility in the start phase.

● "Gel slurry": mix 30 mg of Sephadex IEF with 1 mL of solubilizing mixture.

● Sample application: *sample solution + gel slurry* (1 + 3)

partial volumes

● To separate protein solutions such as serum, plasma etc. the solubilizing mixture is diluted accordingly.

after a quantitative protein determination

● *Tissue proteins* must be solubilized with the solubilizing mixture.

by mechanical disintegration for example

● *Cell proteins* can be extracted by sonication:

Example: yeast cell lysate

● mix 600 mg of lyophilized yeast (Saccharomyces cerevisiae) with 2.5 mL of solubilizing mixture;

● sonicate for 10 min at 0 °C;

● centrifuge for 10 min at 10 °C with 42,000 g;

● mix 10 mL of the supernatant with 30 mL of gel slurry (→ 40 μL), apply 20 μL to the *anode* of the IPG gel.

silver staining concentration

2 Stock solutions

Acrylamide, Bis solution "SEP" (T = 30%, C = 2%):
29.4 g of acrylamide + 0.6 g of Bis, make up to 100 mL with double-distilled water.
Acrylamide, Bis solution "PLAT" (T = 30%, C = 3%):
29.1 g of acrylamide + 0.9 g of Bis, make up to 100 mL with double-distilled water.

*C = 2% in gradient gel solutions prevents the **resolving gel** from peeling off the support film and cracking during drying.*

    Beware! *Acrylamide and Bisacrylamide are toxic in the monomeric form. Avoid skin contact and dispose of the remains ecologically (polymerize the remains with an excess of ammonium persulfate).*

*This solution is used for dilute gels **(IPG) and plateaus**, since they would be unstable at lower crosslinking.*

Gel buffer pH 8.8 (conc 4 ×):

18.18 g of Tris + 0.4 g of SDS , make up to 80 mL with double-distilled water. Titrate to pH 8.8 with 4 mol/L HCl; make up to 100 mL.

Ammonium persulfate solution (APS):

dissolve 400 mg of APS in 1 mL of double-distilled water.

can be stored for one week in the refrigerator at +4 °C

Cathode buffer (conc):

7.6 g of Tris + 36 g of glycine + 2.5 g of SDS,
make up to 250 mL with H_2O_{dist}.

Do not titrate with HCl!

Anode buffer (conc):

7.6 g of Tris + 200 mL of H_2O_{dist}.
Titrate to pH = 8.4 with 4 mol/L HCl;
make up to 250 mL with H_2O_{dist}.

Economy measure: the cathode buffer can also be used here.

0.5 mol/L Tris HCl pH 6.8:

6.06 g of Tris+ 0.4 g of SDS + 80 mL H_2O_{dist}.
titrate to pH 6.8 with 4 mol/L HCl
make up to 100 mL with with H_2O_{dist}.

Dithiothreitol stock solution (2.6 mol/L DTT):
dissolve 250 mg of DTT in 0.5 mL of double-distilled water.

Prepare a fresh solution every day.

Equilibration stock solution: (ESS)

2% SDS	2.0 g
6 mol/L urea	36 g
0.1 mmol/L EDTA	3 mg
0.01% Bromophenol Blue	10 mg
50 mmol/L Tris HCl pH 6.8	10 mL
30% glycerol (v/v)	35 mL of an 87% solution

make up to 100 mL with double-distilled water

Urea and glycerol slow down electro-endosmosis; EDTA inhibits the oxidation of DTT.

● Shortly before the *first equilibration step*, add per *10 mL ESS:*

  200 µL of DTT solution (62 mmol/L)

● Shortly before the *second equilibration step*, add per *10 mL ESS* (neutralization of the excess reducing agent with a fourfold quantity of iodoacetamide (Görg *et al.* 1987):
  200 µL of DTT solution (62 mmol/L)
  + 481 mg of iodoacetamide (269 mmol/L)

Prevention of silver staining artefacts which can come from dust particles etc. The protein pattern is not influenced: Görg A, Postel W, Weser J, Günther S, Strahler JR, Hanash SM, Somerlot L. Electrophoresis. 8 (1987) 122-124.

3 Preparing the gel

IPG strips

The Immobiline gradients pH 4 to 10, 4 to 7, and 7 to 10 are the ones most often employed now for IPG-Dalt separations. Immobiline recipes for other gradients are presented in method 10 (see Tab. 1).

For high resolution IPG-Dalt methods, the IPG gel strips are reconstituted with 8 mol/L urea, 0.5 % CHAPS (w/v), and 10 mmol/L DTT.

Gels for standard separation distances of 10 cm (gel length 110 mm) are cast with the gradient mixer in the normal casting cassette with 7.5 mL each of the dense and the light solutions.

The reswelling cassette is used for long gels, using the gradient maker and blocking the tube. To make an 18 cm long gel, 5.2 mL each of the dense and the light solution are pipetted.

D.S. dense, acid solution, L.S. light, basic solution

Tab. 1: Composition of the polymerization solutions of IPG gels.

IPG gradient stock solutions	pH 4 to 10 D.S.	L.S.	pH 4 to 7 D.S.	L.S.	pH 7 to 10 D.S.	L.S.
Immobiline pK 3.6	551 µL	–	289 µL	151 µL	271 µL	45 µL
pK 4.6	–	57 µL	55 µL	369 µL	–	–
pK 6.2	227 µL	25 µL	225 µL	75 µL	–	–
pK 7.0	45 µL	244 µL	–	135 µL	189 µL	162 µL
pK 8.5	167 µL	78 µL	–	–	175 µL	175 µL
pK 9.3	–	179 µL	–	438 µL	–	140 µL
Acrylamide, Bis "PLAT"	1.0 mL	1.0 mL	1.0 mL	1.0 mL	1.0 mL	1.0 mL
Glycerol (87%)	2.0 mL	0.3 mL	2.0 mL	0.3 mL	2.0 mL	0.3 mL
with H$_2$O$_{Bidist}$ →	7.5 mL	7.5 mL	7.5 mL	7.5 mL	7.5 mL	7.5 mL
TEMED (100 %).	3.5 µL	3.5 µL	3.5 µL	3.5 µL	3.5 µL	3.5 µL

Carefully mix the solutions, measure the pH, then titrate to pH 7 with TEMED or 4 mol/L HCl respectively.

for an optimal and regular polymerization

pipette into the gradient mixer:

- for 11 cm long gels (standard cassette, 25 cm wide)
  7.5 mL of L.S. in the reservoir,
  7.5 mL of D.S. in the mixing chamber,
  + 7 µL of APS (40%) each.

see method 9

- for 18 cm long gels (reswelling cass., 11.5 cm wide, Fig 1):
  5.2 mL of L.S. in the reservoir,
  5.2 mL of D.S. in the mixing chamber,
  + 5 µL of each APS (40%).

Before casting:

Cool the cassette in the refrigerator.

After casting:

Overlayer the gel with distilled water to prevent the penetration of oxygen, leave the cassette for 10 min, so that the gradient can settle horizontally. To facilitate this process add a small volume of glycerol to L.S.

Polymerization: 1 h at 50 °C (heating cabinet).

Washing, drying, rehydrating

- Remove the gel from the casette;

- wash 3 × 20 min in 300 mL of double-distilled water;

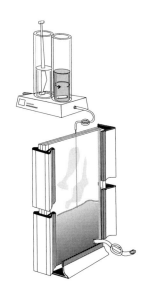

Fig. 1: Casting IPG gels for long separations.

● wash 1 × in 300 mL of 2% glycerol for 30 min;   *so that the gel does not curl up*

● dry at room temperature.   *in a dust-free cabinett.*

● Store wrapped in film at −20 °C (frozen).

    The dried IPG gels are cut into individual strips (110 or   *Mark the strips on the hydro-*
180 × 4 mm) with a paper cutter (see Fig. 2) and left to swell in   *phobic back with a number, and*
a rehydration cassette (see Fig. 3 A and B). The strips should not *mark the cathode and anode*
be broader than 5 mm, in order to minimize the load of nonionic   *with a black marker pen.*
detergent on the SDS gel.

dried gel with
immobilized pH gradient

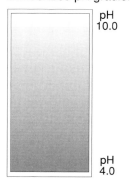

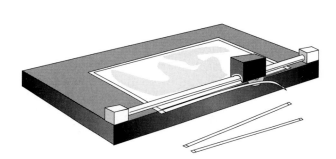

Fig. 2: Precise cutting of the dry IPG strips with a special paper cutter "Roll and cut".

    Vertical rehydration cassette (Fig. 3A): With the urea gels
which are normally used, take the thickness of the support film
into account, this means: cut the GelBond PAG film to the size
of the silicone rubber gasket and place it in the reswelling
cassette. Reswelling of IPG strips *without* urea: *do not* use the
support film frame, otherwise the gels sweat during the separation.

    Horizontal rehydration cassette (Fig. 3B): In such an equip-   *Rabilloud T, Valette, Lawrence*
ment different reswelling solutions can be applied to the strips,   *JJ. Electrophoresis. 15 (1994)*
including reswelling in sample solution as suggested by Rabil-   *1552-1558.*
loud *et al.* (1994). Strips and solutions must be covered with 1   *Volumes per strip:*
to 2 mL paraffin oil during reswelling.   *4 × 110 mm: 275 µL;*
                                                                    4 × 180 mm: 450 µL.

● *Rehydration solution:*

8 mol/L urea	24 g
0.5% CHAPS	250 mg
0.25% (w/v) Pharmalytes pH 3 − 10	310 µL
0.2 % (w/v) DTT	100 mg

*according to the present state of
the art: Görg et al. (1995)*

make up to 50 mL with double-distilled water.

● Rehydration time: 16 h or overnight.

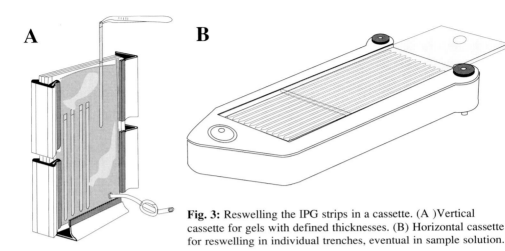

Fig. 3: Reswelling the IPG strips in a cassette. (A)Vertical cassette for gels with defined thicknesses. (B) Horizontal cassette for reswelling in individual trenches, eventual in sample solution.

● *After rehydration:*
sprinkle the surface of the strips with double-distilled water and then place them for a few seconds between two pieces of *wet* filter paper.

so that the urea on the surface does not crystallize out

SDS pore gradient gels

The acrylamide concentrations for the linear pore gradient gels given have been optimized for the separation of yeast cell lysates. The pore gradient can be adapted to the type of separation, which depends on the composition of the sample, by changing the volumes of acrylamide, Bis solution. The sample application plateau contains $T=6\%$ so that it better resists the eventual electro-endosmotic influences of the IPG strips (see Tab. 2).

Here as well, as in method 7, the casting cassette should be cooled and APS added only shortly before pouring, to delay the onset of polymerization.

Tab. 2: Composition of the three gel solutions for a gradient $T = 12\%$ to 15% and a plateau with $T = 6\%$.

"Superdense" plateau does not mix with the gradient.

Stock solutions	Gel solutions		
	Super dense $T = 6\%, C = 3\%$	Dense $T = 12\ \%, C = 2\%$	Light $T = 15\%, C = 2\%$
Glycerol (87 %)	6.5 mL	4.3 mL	–
Acrylamide, Bis"PLAT"	3.0 mL	–	–
Acrylamide, Bis"SEP"	–	6.0 mL	7,5 mL
Gel buffer	3.75 mL	3.75 mL	3.75 mL
Orange G	100 µL	–	–
TEMED	10 µL	10 µL	10 µL
make up to the final volume with double-distilled water	*15 mL*	*15 mL*	*15 mL*

Casting volumes

The volumes of the solutions depend on the gel size desired. There are two possibilities here:

● use the standard gel format as described in method 7.

● use a longer separation distance to obtain a better resolution for highly complex protein mixtures:

gel dimensions: 250 × 190 × 0.5 mm (*"Large scale"*)

see Tab. 3.
APS see Tab. 4

Casting cassettes which are as large as the cooling plate are used for this.

Tab. 3: Gel solution volumes for standard and large scale gels.

Standard gel	V (mL)	Large Scale	V (mL)	pipette into:
Super dense	3.5	Super dense	5.5	→ Cassette
Dense	7.0	Dense	12.5	→ Mixing chamber
Light	7.0	Light	12.5	→ Reservoir

in this case, add APS before: see Tab. 4

Tab. 4: Volumes of APS for standard and large scale gels.

Standard gel	V (µL)	Large Scale	V (µL)	pipette and mix into:
Super dense:	20.0	Super dense:	20.0	→ Test tubes
Dense:	6.0	Dense:	11.0	→ Mixing chamber
Light:	4.0	Light:	7.0	→ Reservoir

and pipette into the cassette

● Add the plateau (super dense) with a pipette or a 20 mL syringe. The gradient is added directly, without polymerization, onto the plateau.

see method 7, SDS-PAGE, do not forget to overlay with 60 % v/v isopropanol-water.

● Let the gel stand for 10 min at room temperature for the density gradient to settle and polymerize at 50 °C for 30 min.

4 Separation conditions

First dimension (IPG-IEF)

IPG focusing in individual strips can be performed with conventional equipment directly on the cooling plate (Fig. 4). To improve sample application and facilitate placing the IPG strips as well as more efficient protection from CO_2 — especially in the basic range of the gradient — it is recommended to use the IPG strip kit (Fig. 4).

In the IPG strip tray the gels can also be run under a layer of silicon or paraffin oil.

IPG strips with conventional equipment:

● Place reconstituted IPG strips beside one another - with the gel side on top - on the cooling plate wetted with kerosene. Place 5 mm wide electrode strips, soaked in double-distilled water, over the end of the IPG strips (Fig. 4);

Use the correct orientation! See method 9.

● Place the silicone rubber frame on the anodic side of the strips;

cut the applicator strips into pieces

● apply 20 µL of each sample (incl. Dextran gel) into each well;

● place the focusing electrodes and connect the cable (Fig. 4).

Make sure that the long anodic connecting cable is plugged in.

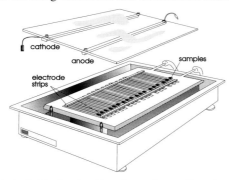

Fig. 4: IEF in individual strips directly on the cooling plate.

IEF with IPG strip kit (Fig.5):

● Place the IPG strips container on the cooling plate coated with kerosene or DC-200 silicone oil and plug in the cable;

● add 1 mL of DC-200 silicone oil to the tank;

do not use kerosene here

● place the reconstituted IPG strips - with the gel side on top - in the grooves;

Make sure the orientation is correct! See method 9.

● place the special electrodes in the contact groove;

● place the sample applicator holder on the anodic side;

● place the sample applicators in the holder in such a way that the cups can be pressed onto the IPG strips;

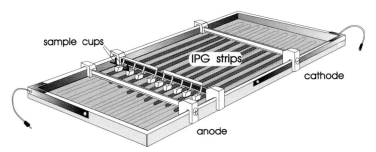

Fig. 5:   IEF in individual IPG strips in the IPG strip kit.

● pipette about 10 mL of DC-200 silicone oil onto the strips; - *Good contact between the cups*
this is not always necessary; *and the gel prevents the oil*
from leaking in.

● pipette 20 μL of each sample (including Dextran gel) into *Sample volumes up to 100 μL*
each well. *can be used.*

Separation conditions

● Set the temperature at 20 °C;

● set the running conditions (Tab. 5). *The voltage must be controlled*
during IEF with IPG strips.

Tab. 5: Power supply settings for IEF.

Current and Power	0.05 mA$_{max}$ and 0.2 W$_{max}$ per strip			
Sample entry	150 V, 30 min 300 V, 1 hour			
Separation	3000 V	3000 V	3000 V	3000 V
Sep. distance	*11 cm*	*11 cm*	*18 cm*	*18 cm*
pH gradient	*4 to 10*	*4 to 7* *7 to 10*	*4 to 10*	*4 to 7* *7 to 10*
Time	5 h	7 h	7 h	Overnight
Volthours	11,000	22,000	18,000	42,000

Note:
20 °C gives better results than
15 °C.
two phases with low voltages in-
dependent of the gel length are
applied for optimized sample en-
try;
according to the present state of
the art: Görg et al. (1995)

Equilibration

Remove the IPG strips from the cooling plate, place them in a
test tube with 10 mL of ESS + DTT and agitate for 15 min. Re-   *see page 218*
peat with 10 mL of ESS + DTT + iodoacetamide. Bend the
equilibrated IPG strips in the shape of a "C" and lay them on   *If the IPG strips are not equili-*
their side on dry filter paper for a minute to remove the excess   *brated immediately, store them*
solvent.   *in liquid nitrogen or at – 80 °C.*

Second dimension (SDS electrophoresis)

Wet the cooling plate with 2 to 3 mL of kerosene or DC-200 silicone oil.

Water is not suitable as it can cause electric shorting.

l   Place the gel onto the center of the cooling plate with the film on the bottom; the side with the stacking gel must be oriented towards the cathode (−). The anodal edge should match exactly with line "5" on the scale on the cooling plate.

wear gloves

l   Lay two of the electrode wicks into the compartments of the PaperPool. If smaller gel portions are used, cut them to size.

*Be sure to use **very** clean wicks, SDS would dissolve any traces of contaminating compounds.*

l   Mix 10 mL of the cathode buffer with 10 mL distilled water and apply it onto the respective strip (Fig. 6).

l   Mix 10 mL of the anode buffer with 10 mL distilled water and apply it onto the respective strip.

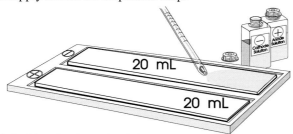

Fig. 6: Soaking the electrode wicks with the respective solution in the PaperPool.

l   Place the cathode wick onto the cathodal edge of the gel; the edge of the wick matching "3.5" on the cooling plate; the anodal wick over the anodal edge, matching "13.5".

Always apply cathode wick first to avoid contamination of catho-de buffer with leading ions.

l   Place the equilibrated and drained IPG strips, with the gel sides to the bottom, on the surface of the SDS gel along the cathodal wick (Fig. 7).

The size of the SDS gels is such that two separations in 11 cm long IPG strips can be run in parallel in a gel.

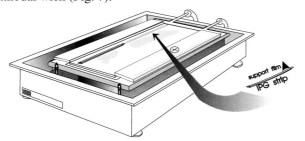

Fig. 7: Placing the equilibrated IPG strip on the SDS gel.

l   Electrophoresis: 15 °C, max. 30 mA, max. 30 W.

l   75 min at max. 200 V, then remove the IPG strips (Fig. 8A)   *Test the IPG strips with Coo-*
     and place the cathodal electrode wick over the IPG-SDS gel   *massie to check whether all the*
     contact area (Fig. 8B).   *proteins have been transferred.*

l   Continue the separation at max. 800 V till the Bromophenol
     Blue has reached the anodic edge of the gel:

Standard gel: 90 min   *Large scale:* 5 h

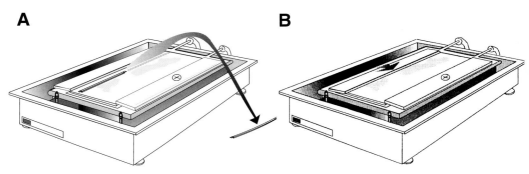

A   **B**

Fig. 8: Removing the IPG strips (A) and shifting the cathodal wicks over the contact surface (B).

5 Coomassie and silver staining

Quick Coomassie staining

l   *Staining:* hot staining by stirring in a stainless steel develo-   *Place the gel with the surface*
     ping tank: 0.02% Coomassie R-350:   *on the grid.*
     dissolve 1 PhastGel Blue tablet in 1.6 L of 10% acetic acid;
     8 min, 50 °C (see method 7);

l   *Destaining:* in the MultiWash in 10% acetic acid for 2 h at
     room temperature (see method 7);

l   *Preserving:* in a solution of 25 mL of glycerol (87% w/v) +
     225 mL of distilled water for 30 min;

l   *Drying:* air-dry (room temperature).

This method is especially adapted to the quantification of spots.

Silver staining

Silver staining is mostly employed, because with a lower protein load less proteins are lost due to aggregation. The most sensitive and reproducible methods are the techniques of Heukeshoven and Dernick (1987),Merril *et al.* (1981), and Blum *et al* (1987). The sensitivity of this methods is of about 0.05 to 0.1 ng/mm^2.

The water and reagents must be very pure and care be taken with cleanliness.

Formaldehyde is a critical reagent: it should not be too old.

A clear background should be obtained with this method.

Tab. 6: Silver staining acc. to Heukeshoven und Dernick.

Step	Solution	V [mL]	t [min]
Fixing	300 mL ethanol 100 mL acetic acid with H$_2$Odist → *1000 mL*	250	>30
Incubation	75 mL ethanol*) 17.00 g Na-acetate 1.25 mL glutaraldehyde (25% w/v) 0.50 g Na$_2$S$_2$O$_3$ × 5 H$_2$O with H$_2$O$_{dist}$ → 250 mL	250	30 or overnight
Washing	H$_2$O$_{dist}$	3 × 250	3 × 5
Silvering	0.5 g AgNO$_3$**) 50 µL formaldehyde (37% w/v) with H$_2$O$_{dist}$ → 250 mL	250	20
Development	7.5 g Na$_2$CO$_3$ 30 µL formaldehyde (37% w/v) with H$_2$O$_{dist}$ → 300 mL if pH >11,5, titrate to this value with NaHCO$_3$ solution	1 × 100 1 × 200	1 3 to 7
Stopping	2.5 g glycine with H$_2$O$_{dist}$ → 250 mL	250	10
Washing	H$_2$O$_{dist}$	3 × 250	3 × 5
Preserving	25 mL glycerol (87% w/v) with H$_2$O$_{dist}$ → 250 mL	250	30
Drying	air-drying (room temperature)		

* First dissolve NaAc in water, then add ethanol. Add the thiosulfate and glutaraldehyde just before use.
** Dissolve AgNO$_3$ in water, add the formaldehyde before use.

Tab. 7: Silver staining according to Merril *et al.* (1981).

Step	Solution	V [mL]	t
Pre-fixing	500 mL methanol 100 mL acetic acid with $H_2O_{dist} \rightarrow 1000$ mL	250	>1 h
Fixing	75 mL ethanol 25 mL acetic acid with $H_2O_{dist} \rightarrow 250$ mL	250	overnight
Post-fixing	150 mL ethanol 50 mL acetic acid with $H_2O_{dist} \rightarrow 500$ mL	2×250	2×10 min
Oxidation	0.6 g $K_2Cr_2O_7$ 600 mL H_2O_{dist} 172 µL HNO_3 (65 %) stir till $K_2Cr_2O_7$ is completely dissolved.	250	20 min
Washing	with H_2O_{dist} 30 µL formaldehyde (37% w/v) with $H_2O_{dist} \rightarrow 300$ mL	4×300	4×30 s
Silvering	0.5 g $AgNO_3$ with $H_2O_{dist} \rightarrow 300$ mL	300	30 min
Pre-rinse	Developer	2×200	$2 \times$ briefly
Development	53.4 g Na_2CO_3 1600 mL H_2O_{dist} 0.9 mL formaldehyde (37 % w/v))	250	15 min
Stopping	2.5 g glycine	250	10 min
Washing	H_2O_{dist}	250	2×10 min
Preserving	25 mL glycerol (87% w/v) with distilled water $\rightarrow 250$ mL	250	15 min
Drying	Air-dry (room temperature)		

This method generates a yellow background when maximum sensitivity is required. Because the development of spots asymptomatically approaches the maximum a very high reproducibility and sensitivity can be obtained.

Evaluation on the light box or by photography.

For separations of plant protein, the method according to Blum *et al.* (1987) has shown to give the highest sensitivity against the lowest background. The following recipe is a modification by Görg *et al.* for gels on film supports:

Blum H, Beier H, Gross HJ. Electrophoresis. 8 (1987) 93-99.

Görg A. personal communication.

Tab. 8: Silver staining according to Blum *et al.* (1987), modified.

Step	Solution	V [mL]	t
Fixing	120 mL methanol 36 mL acetic acid 150 µL formaldehyde (37%) with $H_2O_{dist} \rightarrow$ 300 mL	300	>1 h
Washing	30 % (v/v) ethanol	250	3×20 min
Incubation	50 mg $Na_2S_2O_3 \times 5\ H_2O$ with $H_2O_{dist} \rightarrow$ 250 mL	250	1 min exactly!
Rinsing	H_2O_{dist}	250	3×20 sec exactly!
Silvering	0.5 g $AgNO_3$ 190 mL formaldehyde (37%) with $H_2O_{dist} \rightarrow$ 250 mL	250	20 min
Rinsing	H_2O_{dist}	250	2×20 sec exactly!
Developing	15 g Na_2CO_3 125 µL formaldehyde (37%) 1 mg $Na_2S_2O_3 \times 5\ H_2O$ with $H_2O_{dist} \rightarrow$ 250 mL	250	10 min
Washing	H_2O_{dist}	250	2×2 min
Stopping	125 mL methanol 30 mL acetic acid with $H_2O_{dist} \rightarrow$ 250 mL	250	10 min
Washing	50 % methanol	250	> 20 min
Preserving	25 mL glycerol (87% w/v) with distilled water $\rightarrow$ 250 mL	250	> 15 min
Drying	Air-dry (room temperature)		

The optimal method is dependent on the type of sample; stain three pieces of one gel run!

5 Perspectives

Very basic proteins

For an adequate and reproducible analysis of very basic proteins like lysozyme, histones, and ribosomal proteins in an immobilized pH gradient 9 -12 several modifications of the procedure are necessary (Görg *et al.* 1997, in press):

Görg A, Obermaier C, Boguth G, Csordas A, Diaz J-J, Madjar J-J. Electrophoresis. (1997) in press.

● For casting of the IPG gels, acrylamide has to be replaced by dimethyl-acrylamide.

● 0.2% (v/v) methyl cellulose and 10 % (v/v) isopropanol have to be added to the rehydration solution (s. page 218).

● The first dimension run has to be performed under degassed paraffin oil and continuous flushing with argon gas.

Equilibration of the strips and the running conditions of the second dimension are not modified.

Identification of proteins

Identification of spots and spot groups is performed by further analysis of either the complete polypeptide or of peptide fragments after limited proteolysis.

Protein analysis

After blotting onto the appropriate membrane the complete proteins are identified by

● Amino acid analysis,

● N-terminal sequencing,

● Direct mass analysis with MALDI time of flight mass spectrometry,

and subsequent search in a data base.

Peptide analysis

Proteins are either cleaved in the gel or on the blotting membrane by a tryptic digestion. The peptide fragments are then separated with HPLC or capillary electrophoresis with direct coupling to an ESI quadropole mass spectrometer. Also here a data base is needed for identification.

Method 11: PAGE of double stranded DNA

With the PCR$^{®}$ technique DNA fragments in the size range of 50 to 1,500 bp are amplified. For the analysis of these fragments, the application of thin horizontal polyacrylamide gels on film support with subsequent silver staining show several advantages over conventional agarose gel electrophoresis of DNA fragments:

These gels can be used directly, or after washing, drying and rehydration in the appropriate buffer. The gels can be prepared in the laboratory or purchased ready-made in the dried form.

● Polyacrylamide gels have a higher resolution than agarose gels.

and discontinuous buffer systems can be applied.

● Silver staining has a higher sensitivity (15 pg/band) and is less toxic than ethidium bromide. Silver staining is particularly useful for staining small fragments, because it is independent on the size.

Silver staining does not work well for agarose gels.

● Bands are visible without an UV lamp, no photography is necessary for a permanent record.

Polyacrylamide gels can be dried and put in files.

● Due to the high sensitivity of silver staining, autoradioactivity can be replaced in many cases.

Thus no x-ray films are needed, no radioactive waste.

It should , however, not be forgotten, that the migration in native polyacrylamide gels is not only dependent on the size, but also on the sequence. Thus A and T rich DNA fragments migrate slower than they should according to their sizes.

Example: the "spike" of the 100 bp ladder shows up at 800 bp in agarose gels, but at 1,200 bp in native polyacrylamide gels.

A number of DNA typing methods are performed in gels with the short separation distance: Screening of PCR products, separation of RNA, random amplified polymorphic DNA (RAPD) and DNA amplification fingerprinting (DAF), heteroduplex analysis, and a part of minisatellite analysis.

The gel concentrations can vary from 5 to 15 % T.

In some cases, however, a higher resolution and more space for multiple bands can only be achieved by running gels in the long separation distance: differential display reverse transcription (DDRT) and minisatellite samples with shorter repeats, like the VNTR (variable number of tandem repeat) sample D1S80.

The electrode plate of the Multiphor chamber is square, thus it can be turned 90° and used for long distance separations.

The best results are obtained with discontinuous 0.5 mm thin gels and discontinuous buffer systems. In the following instruction two buffer systems are described: Tris-acetate / Tris-tricine and Tris-phosphate / Tris-borate-EDTA.

Both buffer systems can be employed for directly used gels or for washed gels.

Washed, dried and rehydrated 0.5 mm thin polyacrylamide gels show very high resolution, when rehydrated in the "PhastGel buffer system" (Seymour and Gronau-Czybulka, 1992); and they are free from acrylamide monomers.

Seymour C, Gronau-Czybulka S. Pharmacia Biotech Europe Information Bulletin (1992)

1 Stock solutions

Acrylamide, Bis solution "SEP" (T = 30%, C = 2%):
29.4 g of acrylamide + 0.6 g of bisacrylamide, make up to
100 mL with double-distilled water (H_2O_{Bidist}).
Acrylamide, Bis solution "PLAT" (T = 30%, C= 3%):
29.1 g of acrylamide + 0.9 g of bisacrylamide, make up to
100 mL with double-distilled water (H_2O_{Bidist}).
Caution! *Acrylamide and bisacrylamide are toxic in the monomeric form. Avoid skin contact and dispose of the remains ecologically (polymerize the remains with an excess of APS).*

Ammonium persulfate solution (APS) 40% (w/v):
Dissolve 400 mg of ammonium persulfate in 1 mL double-distilled water.

*C = 2% in the resolving gel so-
lution prevents the* **separation**
*gel from peeling off the support
film and cracking during drying.*

*This solution is used for slightly
concentrated* **plateaus** *with C =*
3%, *because the slot would become unstable if the degree of
polymerization were lower.*

*it can be stored for one week in
the refrigerator (4 °C)*

Buffer System I (Tris-acetate / Tris-tricine)

Gel buffer 0.448 mol/L Tris-acetate pH 6.4 (4 × conc):
5.43 g Tris, dissolve in 80 mL H_2O_{dist}; titrate to pH 6.4 with
acetic acid; make up to 100 mL with H_2O_{dist}.
Anode buffer 0.45 mol/L Tris-acetate pH 8.4:
27.3 g Tris, dissolve in 400 mL H_2O_{dist}; titrate to pH 8.4 with
acetic acid; make up to 500 mL with H_2O_{dist}.
Cathode buffer 0.08 mol/L Tris / 0.8 mol/L tricine:
4.85 g Tris + 71.7 g tricine, make up to 500 mL with H_2O_{dist}.
*Electrode buffer for washed and rehydrated short distance gels
0.2 mol/L Tris-0.2 mol/L tricine - 0.55 % SDS:*
24.2 g Tris + 35.84 g tricine + 5.5 g SDS, make up to 1 L with
H_2O_{dist}.

*With this buffer, the storage
time of the gel is not limited. Because the pH value of the gel is
lower than pH 7, the matrix
does not hydrolyse.*

*For washed gels, electrode solu-
tions with lower concentrations
can be employed (than for non-
washed gels).*

Buffer System II (Tris-phosphate / TBE)

Gel buffer 0.36 mol/L Tris-phosphate pH 8.4 (4 × conc):

4.36 g Tris, dissolve in 80 mL H_2O_{dist}; titrate to pH 8.4 with
phosphoric acid; make up to 100 mL with H_2O_{dist}.
*Electrode buffer 450 mmol/L Tris / 75 mmol/L boric acid /
12.5 mmol/L EDTA-Na2(5 × TBE):*
54.5 g Tris + 23.1 g boric acid + 4.65 g EDTA Na2, make up
to 1 L with H_2O_{dist}.

*The Tris-phosphate buffer is
much better for polymerization
than TBE because borate inhibits the copolymerization of the
gel and the Gelbond PAGfilm.*

Bromophenol blue solution (1%):
100 mg Bromophenol blue, make up to 10 mL with H_2O_{dist}.
Xylencyanol solution (1%):
100 mg Xylencyanol, make up to 10 mL with H_2O_{dist}.
0.2 mol/L EDTA Na2 solution:
7.44 g EDTA Na2, make up to 10 mL with H_2O_{dist}.

Sample buffer:
22 mL H_2O_{dist} + 3 mL gel buffer + 60 µL bromophenol blue solution (1%) + 40 µL Xylencyanol solution (1%) + 250 µL 0.2 mol/L EDTA-Na$_2$ solution.

2 Preparing the gels

Slot former

Sample s are applied in small wells which are molded in the surface of the gel during polymerization. To form these slots a mould must be fixed on a glass plate with a 0.5 mm thin U-shaped silicon rubber gasket.

The cleaned and degreased glass plates with 0.5 mm U-shaped spacer is placed on the template (slot former template in the appendix). For long distance gels the "reswelling cassette" is used. A layer of "Dymo" tape (6 mm wide embossing tape, 250 µm thick) is applied, avoiding air bubbles, at the starting point. The slot former is cut out with a scalpel (see Fig. 1). After pressing the individual slot former pieces against the glass plate, the remains of sticky tape are removed with methanol.

"Dymo" tape with a smooth adhesive surface should be used. Small air bubbles can be enclosed when the adhesive surface is structured, these inhibit polymerization and holes appear around the slots.

The casting mold is then made hydrophobic by spreading a few mL of Repel Silane over the whole slot former with a tissue under the fume hood. When the Repel Silane is dry, the chloride ions which result from the coating are rinsed off with water.

This treatment only needs to be carried out once

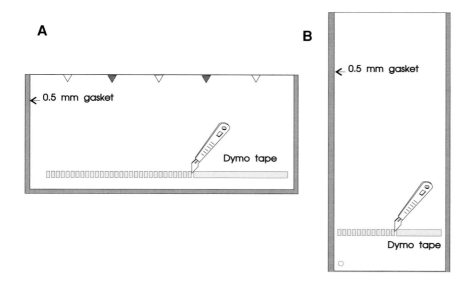

Fig. 1: Preparation of the slot former for short (A) and long (B) separation distances.

Assembling the casting cassette

For mechanical stability and to facilitate handling, the gel is covalently polymerized on a support film. The glass plate is placed on an absorbent tissue and wetted with a few mL of water. The GelBond PAG film is applied with a roller with the untreated hydrophobic side down (Fig. 2). A thin layer of water then forms between the film and the glass plate and holds them together by adhesion. The excess water which runs out is soaked up by the tissue. To facilitate pouring in the gel solution, the film should overlap the long edge of the glass plate by about 1 mm.

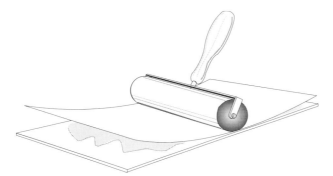

Fig. 2: Applying the support film with a roller.

The finished slot former is placed on the glass plate and the cassette is clamped together (Fig. 3).

The cassette for long gels is clamped along the long side.

Cool the casting cassette in the refrigerator at 4 °C for about 10 min: this delays the onset of polymerization. This step is necessary because the stacking gel with large pores and the resolving gel with small pores are cast in one piece. The polymerization solutions which have different densities take 5 to 10 min to settle.

In the summer in a warm laboratory, the gel solutions should also be brought to 4 °C.

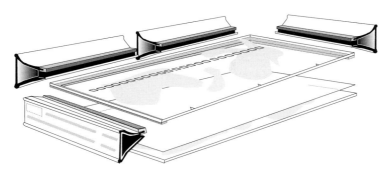

Fig. 3: Assembling the gel cassette for short distances.

Minisatellites (VNTR)

PCR® amplificates are diluted: 1 + 3 parts of sample buffer.
6 µL are applied.
Systems with alleles with repeats of 70 bp (e.g. YNZ 22) can be well resolved in short gels with 10 % T.
16 bp repeats (e.g. D1S80) need long distance gels.

For native PAGE of STRs see Schickle (1996b).

DDRT

For optimized amplification procedures work according to Bosch and Lohmann (1996).
Separations are either performed in native long distance gels or in 15 %T short distance denaturing gels (see method 14).

Bosch TCG, Lohmann J. In Fingerprinting Methods based on PCR. Bova R, Micheli MR, Eds. Springer Verlag Heidelberg, in press.

4 Electrophoresis

In many cases enough resolution is obtained with directly used gels, when double stranded DNA has to be separated in short distances. For long distances and very high resolution it is recommended to use washed and rehydrated gels.

Rehydration of washed and dried gels in the gel buffer

Rehydration solution :

6.25 mL gel buffer + 1 mL glycerol + 1 mL ethylenglycol + 0.5 mL bromophenol blue solution, make up to 25 mL with double-distilled water.

Lay GelPool onto a horizontal table; select the appropriate res-welling chamber, pipet the rehydration solution into the chamber, for

| a complete gel: | 25 mL |
| a half gel (short): | 13 mL |

Set the edge of the gel-film – with the gel surface facing down – into the rehydration buffer (Fig. 5) and slowly lower it, avoiding air bubbles.

The gels for short distances can be used in one piece, or – depending on the number of samples – cut into smaller portions with scissors (when they are still dry). The rest of the gel should be sealed airtight in a plastic bag and stored in a freezer.

Using foreceps, lift the film up to its middle, and lower it again without catching air bubbles, in order to achieve an even distribution of the liquid (Fig. 5 B). Repeat this during the first 10 min.

Very even rehydration is obtained when performing it on a shaker at a slow rotation rate (fig. 5C). If no shaker is used, lift gel edges repeatedly.

60 min later the gel has reswollen completely and is removed from the GelPool. Dry sample wells with clean filter paper, wipe buffer off the gel surface with the edge of a filter paper (Fig. 5D).

When the gel surface is dry enough, this is indicated by a noise like a whistle.

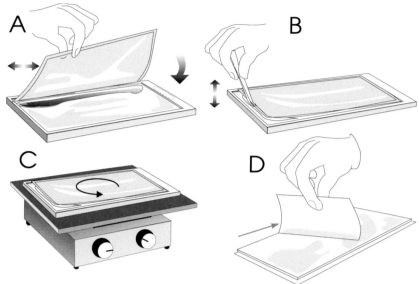

Fig. 5: Rehydration of a gel.
A Placing the dry gel into the GelPool; **B** Lifting the gel for an even distribution of the liquid.
C Rehydration on a rocking platform (not always necessary). **D** Removing the excess buffer
from the gel surface with filter paper.

Preparation of the electrode wicks

Short distance gels: Lay two of the 25 × 5 cm electrode wicks
into the compartments of the PaperPool. Apply 20 mL of the
respective electrode buffer to each wick (Fig. 6).

Do not forget, that for one buffer system different anode and cathode buffer is used.

Long distance gels: Cut six electrode strips of 11.7 ×1.8 cm. Lay
three of them stacked into the compartments of the PaperPool.
Apply 10.5 mL of the respective buffer to each stack.

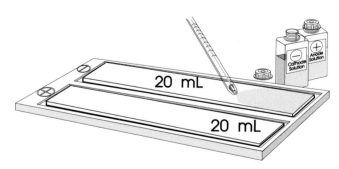

Fig. 6: Soaking the wicks with electrode buffer.

Application of the gel and the electrode wicks

Switch on the thermostatic circulator, adjusted to the 15 °C. Apply a very thin layer of kerosene (ca. 0.5 mL) onto the cooling plate with a tissue paper, in order to ensure good cooling contact (a few air bubbles do not matter).

Short distance run: Place the gel (surface up) on to the center of the cooling plate: the side containing the wells must be oriented towards the cathode (Fig. 7; Multiphor: line 5).

Place the cathodal strip onto the cathodal edge of the gel, edge of the electrode strip matching "3.5" on the cooling plate. The edge of the strip should be at least 4 mm away from the edges of the sample wells (otherwise small DNA fragments will exhibit less sharpness or hollow bands).

Place the anode strip over the anodal edge, matching "13.5" on the cooling plate. Smooth out air bubbles by sliding bent tip forceps along the edges of the wicks laying in contact with the gel.

Long distance run: Place the gel (surface up) on to the center of the cooling plate: the side containing the wells must be oriented towards the cathode .

Place the cathodal stack electrode strips to the cathodal side of the gel, the inner edge matching with line "23" of the cooling plate.

Place the other stack over the anodal side, the inner edge matching with line "2" of the cooling plate. Smooth out air bubbles by sliding bent tip forceps along the edges of the wicks laying in contact with the gel.

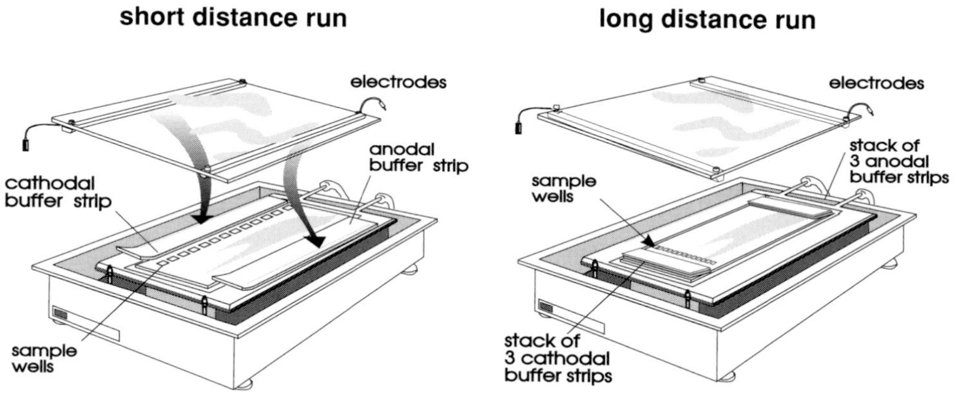

Fig. 7: Appliance for short and long distance runs of DNA fragments.

Sample application and electrophoresis

Apply 6 μL of each sample to the sample wells using a micropipette. Clean platinum electrode wires before (and after) each electrophoresis run with a wet tissue paper. Move electrodes so that they will rest on the outer edges of the electrode wicks. Connect the cables of the electrodes to the apparatus and lower the electrode holder plate (Fig 7). Close the safety lid.

Running Conditions

The suggested running conditions have to be modified for some applications: e.g. when samples of the size range 700 to 900 bp have to be well separated, at least two hours separation time is required. The dyes added to the samples are a help for the estimation of the runnning time:

In a 10% *T* gel the Xylencyanol dye migrates with the same mobility like 200 bp fragments, in a 15 %*T* gel it migrates like 100 bp fragments.

Short distance run in a 10 % T gel,
Tris-acetate / tris-tricine buffer, 15 °C, whole gel:

600 V_{max},	25 mA_{max}	15 W_{max},	1 h 20 min

Short distance run in a 10 % T gel,
Tris-phosphate / tris-borate buffer, 15 °C, whole gel:

100 V_{max},	10 mA_{max}	5 W_{max},	20 min
600 V_{max},	30 mA_{max}	10 W_{max},	45 min

Long distance runs:

Note: High voltage and relatively low milliampere values are applied on long electrophoresis gels. This means, that with a milliampere-constant control only little changes in conductivity (e.g. from 25 to 28 mA) can result in severe changes of the voltage value, e.g. from 250 to 360 V.

The electrophoresis pattern in long gels is highly dependent on the temperature and the running conditions. The conductivities of the cables and electrodes in the electrophoresis chamber have a great influence on the separation.

Different chambers have different cable and electrode lengths and diameters, which results in different conductivities inside the apparatus.

The following power supply settings are valid for one chamber type. They may be changed for a different chamber. The actual **voltage** values are most important for optimized separation patterns.

Long distance run in a 10 % T gel,
Tris-phosphate / tris-borate buffer, 15 °C:

Phase	set	actual	set	actual	set	actual	time
1	150 V_{max}	*150 V*	15 mA_{max}	*11 mA*	5 W_{max}	*2 W*	30 min
2	400 V_{max}	*400V*	31 mA_{max}	*31 mA*	10 W_{max}	*10 W*	10 min
3	500 V_{max}	*360V*	24 mA_{max}	*24 mA*	10 W_{max}	*7 W*	45 min
4	800 V_{max}	*490V*	28 mA_{max}	*28 mA*	13 W_{max}	*13 W*	2 h 20 min
total:							***3 h 45 min***

Control with Xylencyanol band: m_R= 230 bp should be ca. 1 cm away from the edge of the anodal filter paper.

When a different system is employed, the mA values the 3rd and 4th phase have to be adjusted to reach the actual voltage values of 360 V and 490 V, respectively, at the beginning of these phases: for example 3 → 22 mA_{max} and 4 → 23 mA_{max} .

Voltage-Ramping:

With a programmable power supply the voltage values can be controlled with a program. If the current and power is set to a value, which can never be reached during the run, the power supply can control the voltage over the time: "Voltage Ramping".

The method used in the power supply has to be switched in the menu "Setup" from

● Volt Level=fixed   to

● Volt Level=changing or =gradient)

The power supply calculates the voltage values over the time value. At least two phases have to be defined. The power supply calculates a linear transition of the voltage from phase to phase (gradient!). The timing of the voltage changes is now fixed.

Further advantages of this method:

- If additives (e.g. urea, glycerol) are present in the gel, or the buffer system is slightly varied (pH value, concentration), the settings of the power supply do not have to be altered.

- The width of the gel does not matter.

RAMPING programme

Voltlevel: gradient:

set	set	set	time
150 V_{max}	30 mA_{max}	20 W_{max}	1 min
150 V_{max}	30 mA_{max}	20 W_{max}	29 min
400 V_{max}	50 mA_{max}	30 W_{max}	1 min
400 V_{max}	50 mA_{max}	30 W_{max}	29 min
500 V_{max}	50 mA_{max}	30 W_{max}	45 min
800 V_{max}	50 mA_{max}	30 W_{max}	2 h 20 min
total:			***3 h 45 min***

5 Silver staining

Tab. 2: Sensitive silver staining acc. to Bassam *et al.* (1991).

Bassam BJ, Caetano-Annollés G, Gresshoff PM. Anal Biochem (1991) 81-84.

1. Fix	30 min	250 mL	10 % acetic acid (v/v)
2. Wash	3 × 5 min	3 × 250 mL	H_2O_{Bidist}
3. Silver	*20 min*	*200 mL* + 200 µL formaldehyde	*0.1 % $AgNO_3$ (w/v)*
4. Rinse	20 s	250 mL	H_2O_{Bidist}

Thoroughly wash gel surface and back with a squeeze bottle.

5. Develop	2 – 5 min (visual control)	200 mL	2.5 % Na_2CO_3 + 200 µL formaldehyde + 200 µL Na-thiosulphate (2 %)
6. Stop / *Desilver*	*10 min*	*250 mL*	2.0 % (g/v) glycine + 0.5 % EDTA Na_2 solution
7. Impregnate	10 min	250 mL	5 % glycerol (v/v) for 10 % gels 10 % glycerol (v/v) for 15 % gels
8. Air dry		*room temperature*	

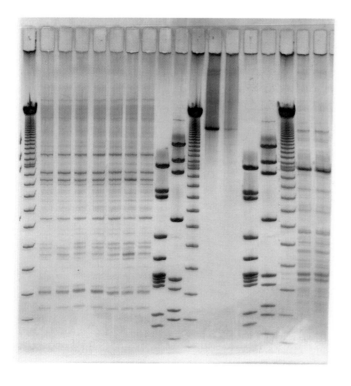

Fig. 8: Separation of DNA fragments in a rehydrated polyacrylamide gel with a discontinuous buffer system. Silver staining.

Tab. 3: Quick silver staining acc. to Budowle et al. (1991).

1. Fix / Oxidise	3 min	100 mL	1 % nitric acid (v/v) until bromophenol blue has turned into yellow
2. Rinse	5 sec	200 mL	H₂O_Bidist
3. Silverg	20 min	200 mL	0.2 % AgNO₃ (w/v)
4. Rinse	5 s	200 mL	H₂O_Bidist
5. Develop	2 – 5 min (visual control)	200 mL	3 % Na₂CO₃ + 170 μL formaldehyde

Budowle B, Giusti AM, Waye JS, Baechtel FS, Fourney RM, Adams DE, Presley LA et al. Am J Hum Genet. 48 (1991) 841-855.

Changing the development solution may be necessary; the solution should be changedd when it turns brown.

6. Stop	10 min	250 mL	10 % (v/v) acetic acid
7. Impregnate	10 min	250 mL	5 % glycerol (v/v)for 10 % gels 10 % glycerol (v/v) for 15 % gels
8. Air dry			room temperature

Tab. 4: Very rapid silver staining acc. to Sanguinetti et. al (1994).

Sanguinetti CJ, Dias NE, Simpson AJG. BioTechniques. 17 (1994) 915-919.

1. Fix	3 min	100 mL	10 % ethanol + 5 % acetic acid (v/v)
2. Silver	5 min	200 mL	10 % ethanol + 5 % acetic acid (v/v) + 0.2 % AgNO₃ (w/v)
3. Rinse	20 s	200 mL	H₂O_Bidist
4. Rinse	2 min	200 mL	H₂O_Bidist
5. Develop	2 – 5 min (visual control)	200 mL	2 % NaOH 0.5 % formaldehyde
6. Stop	5 min	250 mL	10 % ethanol + 5 % acetic acid (v/v)

The three staining procedures show different sensitivities:

Method	Duration	Sensitivity
(Bassam)	1 hour	100 pg / band
(Budowle)	30 min	300 pg / band
(Sanguinetti)	15 min	500 pg/ band

All three protocols are adequate for reamplification of stained DNA fragments.

Reamplification is not possible, when a protocol for silver staining of proteins has been employed.

Ethidium bromide staining can not be used, because, the polester support film is fluorescent. However, fluorescent labels for red light, like Cy5 can be employed.

Method 12: Native PAGE of single stranded DNA

The most certain and sensitive method for the detection of mutations is DNA sequencing analysis. However, this method is too costly and time-consuming for screening purposes.

SSCP (single strand conformation polymorphism) analysis is a very powerful technique for mutation detection, and it can easily be applied for screening purposes.

The principle: Variations in the sequence as small as one base exchange alter the secondary structure of ssDNA. Changes in the sequence cause differences in the electrophoretic mobility, which are observed as band shifts (Orita *et al*, 1989). The mechanism of SSCP is described as: Differential transient interactions of the bent and curved molecules with the gel fibers during electrophoresis, causing the various sequence isomers to migrate with different mobilities.

Figure 1 shows a SSCP analysis gel with silver stained ssDNA fragments of p53 genes (Schickle, 1996a).

Before screening, the mutants have to be defined by direct sequencing. PCR products are denatured by heating with formamide or sodium hydroxid, chilled in an ice-water bath, and loaded onto a non-denaturing polyacrylamide gel for electrophoresis.

A wide range of new methods in clinical diagnostics is based on mutation detection.

As already mentioned on page 24, SSCP can also be employed for genetic differentiations.

The folded structure of the single strands is maintained by sequence-specific intramolecular base-pairing, which can be influenced by the running conditions.

Schickle HP. GIT LaborMedizin. 19 (1996a) 159-163.

The sequences of the two primers for the amplified fragment have to be detected.

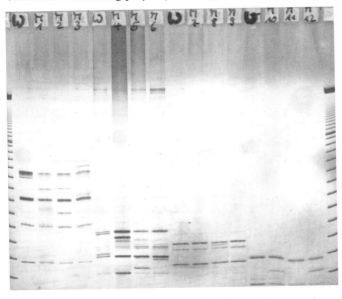

Fig. 1: SSCP analysis: Different exons in p53 mutation screening. Horizontal 0.5 mm thin polyacrylamide gel 10 % *T*, 2 % *C*. 10 °C, silver staining; for the other conditions see Schickle HP (1996a).

Many samples can be screened with a considerably lower effort than direct sequencing in a relatively short time, namely within a few hours.

SSCP analysis can be performed in all types of electrophoresis chambers for polyacrylamide gels, provided that the temperature can be controlled. However, horizontal systems have several advantages over vertical systems:

● With Peltier cooling the temperature can be exactly controlled.

● The gel size can be easily modified.

● Washed and rehydrated gels can be employed.

● Experiments for the optimization of conditions are easy to perform.

● The handling of film-supported gels is very easy.

Silver staining of the DNA fragments can replace the radioisotopic detection methods, when this multi-step procedure can be performed without breaking of the gel. Staining of large and thin, unbacked gels, which have been run in vertical chambers are not easy to stain and handle.

Ethidium bromide can not detect single stranded DNA.

SSCP – like other DNA analysis methods – can not distinguish between silent mutations which do not cause a disease, and mutations that are disease-associated.

This can be checked on the RNA or protein level.

SSCP analysis is not a replacement but an addition to sequencing, when 100 % of defined mutations have to be detected. The band shifts do not show up automatically for all mutations and under all conditions. Unfortunately, there is not a single and unique separation condition, which can be applied to the separations of all exons.

For some exons only 60 % of mutations can be detected so far.

Several parameters influence the results in SSCP analysis:

Hayashi and Yandell (1993); Jaksch M, Gerbitz K-D ,Kilger C. J Clin Biochem. 28 (1995) in press

DNA-fragment-length,
denaturation procedure,
gel concentration,
crosslinking factor,
buffer system,
gel length,
temperature,
field strength,
running time,
additives (e.g. glycerol).

Thus, when a mutation does not appear in the first experiment, that does not mean, that SSCP analysis can not be employed for this exon.

The following chapter should be seen as a support for selecting the optimal media and conditions for SSCP analysis.

1 Sample treatment

Fragment size

The fragments should not be longer than 200 bp. The mutation should be closer to the center of the amplified sequence than to the end.

Denaturing

In order to find the best and most reliable denaturation procedure, the author has randomly checked publications on SSCP analysis, which are listed in table 1. It is sure, that the way of denaturation has a great influence on the conformation of the singel strands. Also the effectiveness of chilling, and the time for pipetting on the gel play an important role.

The author has stopped after the eighth publication.

The formamide solution should always contain 0.05% bromophenol blue and 0.05 % xylencyanol for easy pipetting and migration control.

This solution is sometimes called "stop solution".

Tab. 1: Denaturating procedures for SSCP analysis from eight randomly selected publications.

PCR product	formamide	EDTA mmol/L	NaOH mmol/L	temp °C	time min	
20 µL	–	–	–	95	5	
10 µL	10 µL	–	–	85	3	
10 µL	10 µL	–	–	95	3	
5 µL	10 µL	–	–	95	5	
3.5 µL	5.5 µL	–	–	95	5	
20 µL	–	1	33	50	10	*No formamide is added, because this would lead to additional heteroduplex band patterns.*
1 µL	9.0 µL	20	–	95	2	
10 µL	10 µL + 4.6 mol/L urea	5	–	94	10	*A three step procedure!*
				65	5	
				20	15	

Chilling of the samples must be performed in ice-water, because the cooling effect of crushed ice is insufficient.

Pipetting time should be as short as possible. If there are samples with a strong tendency to form double strands and a high number of samples has to be pipetted, multi-channel pipettes can be useful.

2 Gel properties

Composition

Vidal-Puig and Moller (1994) have compared different gel media with different exons, and could show differences in the sensitivity for mutation detection.

Vidal-Puig A, Moller DE. Bio-Techniques 17 (1994) 492-496.

The total monomer concentration can vary from 5 to 20 %T. With a stacking gel the resolution is improved.

In general, the lower the hydrophobicity of the matrix, the higher is the sensitivity of mutation detection. This hypothesis could be proved by the discovery, that more mutations were detected in gels with low crosslinking. The gel compositions described methods 1, 4, 7 and 11 are suitable for SSCP analysis, because the resolving gels have a crosslinking factor of only 2 %C.

A gel with high crosslinking factor is more hydrophobic than a low crosslinked gel.

Also ready-made gels with 2% crosslinking are available.

Because temperature is an important parameter, it should be kept in mind, that the Joule heat during electrophoresis is developed by all ionic conmpounds inside the gel. The use of washed, dried and rehydrated gels is recommended.

For preparation and rehydration of these washed and dried gels, see methods 4 and 11.

Gel length

It has been shown in many publications, that a separation distance as short as 4 cm can be sufficient. However, minor mobility differences are better resolved in larger gels.

3 Buffers and additives

Buffer composition

SSCP analysis in vertical gels is mainly run in TBE (tris-borate EDTA) buffer. However, the composition of TBE buffer is not standardized: In the literature the content of boric acid varies from 75 to 90 mmol/L, the EDTA from 2.5 to 20 mmol/L. Good results are also achieved in SDS gels with the standard buffer compositions.

For *vertical gels* the following buffer according to Maniatis *et al.* (1982) is suggested:

For the recipe see method 15.

TBE system

90 mmol/L Tris / 75 mmol/L boric acid / 12.5 mmol/L EDTA-Na2

in both, the gel and the electrode tanks.

For *horizontal gels* with electrode strips the following disconti-nuous buffers produce good results:

For the recipes see method 11.

Tris-acetate / Tris-tricine system

Gel: 112 mmol/L Tris-acetate pH 6.4,
               3.75 % glycerol, 3.75 ethylenglycol.
Anode: 0.45 mol/L Tris-acetate pH 8.4,
Cathode: 0.08 mol/L Tris / 0.8 mol/L tricine.
Alternative electrode buffer for washed and rehydrated or very short gels (PhastSystem[®]):
              0.2 mol/L Tris-0.2 mol/L tricine - 0.55 % SDS

With this buffer, the storage time of the gel is not limited. Because the pH value of the gel is lower than pH 7, the matrix does not hydrolyse.

Tris-phosphate / TBE system

Gel: 90 mmol/L Tris-phosphate pH 8.4 ,
              3.75 % glycerol, 3.75 ethylenglycol.
Electrodes: 450 mmol/L Tris / 375 mmol/L boric acid /
              12.5 mmol/L EDTA-Na2 (5 × TBE)

Additives

At least 10 % glycerol should be added to the rehydration or polymerisation solution. It is disputed, whether SDS is helpful or not.

4 Conditions for electrophoresis

Temperatures of 5, 15 or 25 °C produce different mobility shifts for single stranded DNA. Some laboratories run the same sample three times at these different temperatures to be sure not to miss a mutation.

The gel temperature must be exactly controlled.

The field strength has to be limited to less than 20 V/cm for some samples. As a consequence, some experiments take four hours up to over night.

In some cases the field strength does obviously not cause migra-tion differences.

As already mentioned above, care should be taken, that the chilling and pipetting time does not become too long.

5 Strategy for SSCP analysis

● Use short fragments (< 200 bp).

● Denature samples by heating for 3 minutes at 95 °C with 50 % formamide (1 part PCR product + 1 part formamide) and place them in ice-water for at least 5 min prior to application on the gel.

● Try the most convenient way: ready-made gels and buffers, or the gels and buffers described above, containing 10 % (v/v) glycerol.

Even standard SDS gels, as described in methods 7 and 15 can be employed.

● The resolving gels must not have more than 2 % crosslinking.

● Try 15 °C.

● Apply the standard field strength suggested for the gels and buffers used.

See methods 7, 11, and 15 for running conditons.

If there is a problem of the single strands forming double strands during sample application, it can be tried to solve the problem by adding formamide to the stacking gel. Either cast a gel in this way or equilibrate a film-supported gel in a buffer solution containing 40 % (v/v) formamide like described in method 7 on page 187.

When, after the first trial, the defined mutant(s) is (are) not distinguishable from the wildtype, try the following modifications in the suggested sequence:

● Try different temperatures, first 5 °C, then 25 °C.

● Use low field strength (20 V/cm) for as long time as possible on the equipment available.

● Try higher gel concentration: 15 or 20 %T.

● Try different buffers in rehydrated gels*).

● Add more glycerol or other hydrophilic additives like monoethylenglycol*).

Here are chances for new developments.

● Modify denaturation procedure (see table 1).

● Modify fragment length, try another primer pair.

● Try DGGE, TGGE or other mutation detection methods.

*) One of the benefits of washed and dried gels on film support is, that they can be cut into strips, the strips can be rehydrated in different buffers and / or additives, and they can be run side by side on the cooling plate.

Method 13: Denaturing gradient gel electrophoresis

With denaturing gradient gel electrophoresis (DGGE) single base exchanges in segments of DNA can be detected with almost 100 % efficiency. The principle of DGGE is based on the different electrophoretic mobilities of partially denatured molecules caused by differences in DNA melting.

This procedure needs much more efforts than SSCP, and is therefore by far less frequently employed.

Typically the 100 % denaturant gel part contains 7 mol/L urea and 40 % formamide. The gels are run at temperatures between 40 ° and 60 °C.

A denaturant gradient perpendicular to the electrophoresis direction allows the identification of the region of a point mutation. A denaturant gradient parallel to the electrophoresis run is better for screening, because many samples are run in one gel. Constant denaturing gel electrophoresis (CDGE) is employed for screening, when the denaturant concentration of differential melting of a DNA segment has been detected with DGGE.

Also with these techniques it is not possible to distinguish between silent mutations which do not cause a disease, and disease-associated mutations.

In practice, there are different technical solutions of preparing the gels and running them:

● Freshly prepared gradient gels are used in a vertical or a horizontal system.

The vertical system must have a special design because of the heating.

● Washed and dried gels are rehydrated in a gradient and run horizontally.

The latter method uses a technique, which has been introduced by Altland *et al.* (1984) for preparing immobilized pH gradient gels with a perpendicular urea gradient. Barros *et al.* (1991) have rehydrated home-made prepolymerized, washed and dried gels in a denaturant gradient, with 6 mol/L urea, 20 % formamide, and 25 % glycerol in the 100 % denaturant solution for DGGE, and run them on a horizontal chamber, the PhastSystem.

In the following instruction very simple and reproducible procedures are described for preparing and running DGGE in perpendicular and parallel gradients, as well as CDGE. For the preparation of the washed and dried gels follow the instructions in method 11.

1 Sample preparation

In the fragments analysed, the mutation must be in the part of the sequence with a low melting domaine. In order to detect every mutation, a GC clamp with 30 to 40 bp is frequently added to the fragment during PCR as a high melting domain. The fragment sizes are between 100 and 900 bp.

2 Rehydration solutions

Dye: Prepare a 1 % (w/v) stock solution of bromophenol blue.

Gel buffer 0.36 mol/L Tris-phosphate pH 8.4 (4 × conc):

4.36 g Tris, dissolve in 80 mL H_2O_{dist}; titrate to pH 8.4 with phosphoric acid; make up to 100 mL with H_2O_{dist}.

Electrode buffer 450 mmol/L Tris / 75 mmol/L boric acid / 12.5 mmol/L EDTA-Na2(5 × TBE):

54.5 g Tris + 23.1 g boric acid + 4.65 g EDTA Na2,   make up to 1 L with H_2O_{dist}.

Tab. 1: Rehydration solutions: dense denaturing (7 mol/L urea and 40 % (w/v) formamide) and light native.

	Dense denat.	Light native
urea	4.2 g	–
gel buffer (4×conc)	2.5 mL	2.5 ml
formamide	4.0 mL	–
bromophenol blue (1%)	0.2 mL	–
distilled water	–	7.5 mL
mix thoroughly	=10 mL	= 10 mL

3 Preparing the rehydration cassette

a) Perpendicular gradient

Two glass plates 12.5 ×12.5 cm are needed. Cut a washed and dried gel in the middle (Fig. 1A) . Place one glass plate on a paper towel. Pipette ca 3 mL water on the surface. Lay the gel with the carrier film downward on the liquid, which distributes evenly between glass and film. The gel-film must jut out with the edge, which has no film margin.

Preferably a 15 %T / 2 % C gel is used here, which is prepared according to the instructions in method 11 and 14.

Tilt the glass plate with the film up and let it stand on the paper towel for ca. 5 min (Fig. 1B). The excess water will be drawn out by gravitation and soaked off by the paper: in this way the film will adhere firmly and evenly to the glass plate.

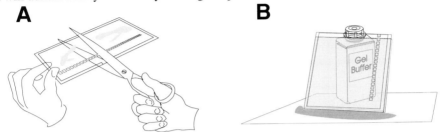

Fig. 1: A: Cutting the dried gel to size. B: Generation of a thin water film to attach the hydrophobic side of the carrier film evenly to the glass plate.

Prepare an U-shaped gasket: Using a scalpel, cut a square of 12.5 × 12.5 cm out of a 0.5 mm silicon rubber plate or a stack of four layers of Parafilm®. Cut a rectangle of 11.5 × 12.0 cm out of this square in that way, tha an U-shaped gasket of 5 mm width is left.

Unfortunately this gasket it not available as a ready-made product.

The glass plate with the attached dry gel-film is placed on the bench horizontally, and the gasket is laid on the three margin edges of the dry gel-film (Fig. 2). Then the second glass plate is laid on top; the sandwich is clamped together with the clamps, using two clamps at the bottom (see Figs. 3 and 4).

The gel-film should not be attached to the glass plate with a roller, because the surface is sticky and sensitive to scratching.

Fig. 2: Assembling the rehydration cassette.

b) Parallel gradient

Complete gel: The large glass plate and the U-frame is used.   *see method 11*

Half gel: Cut the washed and dried gel on all sides to 125 mm. The sample wells must be oriented towards the upper part of the cassette (see Fig. 4). The gel-film is laid on the glass plate with the upper edge protruding by ca. 1 mm to have a lead for the tip of the tubing of the gradient maker. All other steps are performed like described above (see also Fig. 2).

c) Constant denaturing gels

Complete gel: The large glass plate and the U-frame is used.   *see method 11*
Half gel: Cut the washed and dried gel on all sides to 125 mm.
The sample wells must be oriented towards the upper part of the
cassette (see Fig. 4). The gel-film is laid on the glass plate with
the upper edge protruding by ca. 1 mm to have a lead for the tip
of the tubing of the gradient maker. All other steps are performed
like described above (see also Figs. 1 and 2).

4 Rehydration

The perpendicular gradient

● The valve and the clamp on the tubing must be closed.

● Pipet **4 mL** of the light solution into the reservoir. The
   compensation bar is placed into the reservoir. The valve is
   shortly opened to fill the connection between reservoir and
   mixing chamber.

● The magnetic bar is put into the mixing chamber. Then **4 mL**
   of the dense solution are pipetted into the mixing chamber.

● The gradient maker is placed on a magnetic stirrer motor, the
   outlet 5 cm above the edge of the cassette. The magnetic
   stirrer is started with a speed to give only a light vortex.

● The tip of the tube is placed in the middle of the cassette.   *See figure 3.*

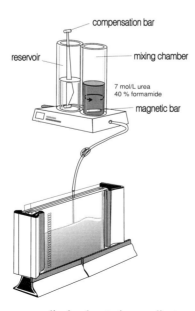

Fig. 3: Casting a perpendicular denaturing gradient.

● First the valve of the gradient maker is opened, secondly the clamp on the tubing.

● Now the gradient is flowing into the chamber.

If it does not flow (mostly because of remains of water in the tubing from last washing), close the gradient maker valve, push a thumb or a finger into the mixing chamber, open the valve after the liquid has started to flow.

● The gradient maker must be cleaned and dried before next use.

● The cassette is left for rehydration for 2 hours.

The parallel gradient

*(volumes are given for a complete gel, **divide by 2** for a half gel)*

● The valve and the clamp on the tubing must be closed.

● Pipet **6 mL** of the light solution into the reservoir. The compensation bar is placed into the reservoir. The valve is shortly opend to fill the connection between reservoir and mixing chamber.

● The magnetic bar is put into the mixing chamber. Then **6 mL** of the dense solution are pipetted into the mixing chamber. The magnetic bar is put into the mixing chamber.

● The gradient maker is placed on a magnetic stirring motor, the outlet 5 cm above the edge of the cassette. The magnetic stirrer is started with a speed to give only a light vortex.

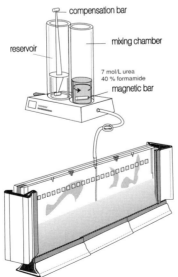

Fig. 4: Casting a parallel denaturing gradient

- The tip of the tube is placed in the middle of the cassette. See figure 4.
- First the valve of the gradient maker is opened, secondly the clamp on the tubing.
- Now the gradient is flowing into the chamber.

  If it does not flow, see perpendicular gradient.
- When the gradient maker is empty, pipette **4 mL** of light solution into the cassette (the stacking gel zone should not contain a gradient!).
- The gradient maker must be cleaned and dried before the next use.
- The cassette is left for rehydration for 2 hours.

Constant denaturing gel

- Mix dense and light solution in a certain percentage according to the result of a perpendicular DGGE experiment.
- **Complete gel**: Pipette **16 mL** of this solution into the cassette.
- **Half gel:** Pipette **8 mL** of this solution into the cassette.
- The cassette is left for rehydration for 2 hours.
- Alternatively, rehydration of a constant denaturing gel can be performed in an horizontal tray (GelPool). See figure 5.

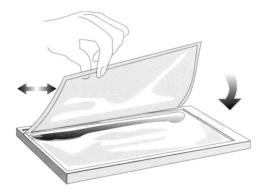

Fig. 5: Rehydration of a dried gel in the GelPool.

5 Electrophoresis

After rehydration, the cassette is dismantled, the gel is laid on a horizontal plane, with the edges of a filter paper the excess of solution is soaked off the surface (Fig. 6) - always from the nondenaturing towards the denaturing side!

Drying is performed until you can hear a squeaking.

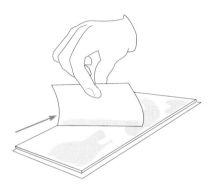

Fig. 6: Removing the excess buffer with the edge of a clean filter paper.

- Switch on the thermostatic circulator, adjusted to 55 °C. Apply a very thin layer of kerosene (ca. 0.5 mL) onto the thermostatic plate with a tissue paper, in order to ensure good heating contact (small air bubbles do not matter). Place the gel (surface up) on to the center of the cooling plate: the side containing the wells must be oriented towards the cathode. Two half gels can be run together on the chamber shown in figure 8.
- Lay two of the electrode strips into the compartments of the PaperPool. Apply 20 mL (10 mL for half strips) of the electrode buffer to each of the wicks (Fig. 7).

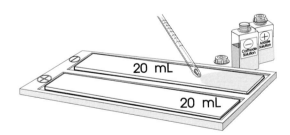

Fig. 7: Soaking the filter paper strips with electrode solutions.

- Place one of the strips onto the cathodal edge of the gel, edge of the electrode strip matching line 3.5 on the cooling plate. Place the other strip over the anodal edge, matching "13.5" on the cooling plate . Smooth out air bubbles by sliding bent tip forceps along the edges of the wicks laying in contact with the gel.
- Apply 6 µL of each sample to the sample wells using a micropipette.

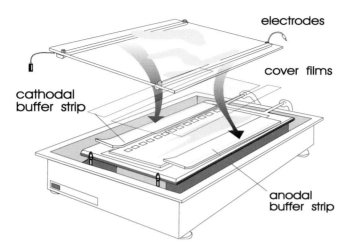

Fig. 8: Appliance for DGGE run at high temperatures. Note the cover films, which are applied after the samples have entered the gel.

● Move electrodes so that they will rest on the outer edges of the electrode wicks.

● Connect the cables of the electrodes to the apparatus and lower the electrode holder plate. Close the safety lid.

Running conditions (55 °C) for two half gels or one complete gel:

| 100 V_{max} | 15 mA_{max} | 5 W_{max} | 20 min |
| 600 V_{max} | 28 mA_{max} | 10 W_{max} | 1 h 30 min |

● After 20 minutes, the first phase, stop the run, cover the gel surface and the electrode strips – between the electrode wires – with a polyester film, to prevent them from drying out and continue the separation.

● Clean platinum electrode wires after each electrophoresis run with wet tissue paper.

6 Silver staining

It is suggested to employ the most sensitive silver stainnig protocol as described on page 240 with one exception: The fixation has to be modified for urea containing gels:

*Fix:*30 min250 mL 15 % ethanol + 5 % acetic acid (v/v) at + 40 °C with the gel swimming on the surface of the liquid, film-side up.

Method 14: Denaturing PAGE of DNA

Denaturing polyacrylamide gels are employed for the following techniques:
- DNA sequence analysis
- Amplified restriction fragment length polymorphism (AFLP)
- Microsatellite analysis
- Differential display reverse transcription (DDRT)
- Ribonuclease protection assay (RPA)
- Temperature gradient gel electrophoresis (TGGE)

Most laboratories perform all these techniques – except the TGGE – in a sequencing apparatus.

In many cases it is not necessary, to run long sequencing gels and use radioactivity for detection. In figure 1 a silver stained 15%T gel is shown, which has been washed, dried and rehydrated in urea-buffer.

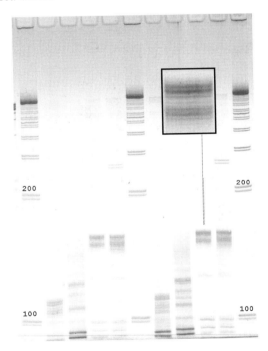

Fig. 1: Denaturing electrophoresis of microsatellites (PCR products of cancer tissues). Silver Staining. The bands in the enlargement show the resolution of one base.

When the samples are run on thin horizontal gels polymerized on carrier films, a number of advantages is achieved:

● The gels are much easier to handle.

● Washed and dried gels can be freshly rehydrated with urea-buffer.

● The gels can be silver stained.

DNA bands can be reamplified.

● When radioactivity is needed, the amount of contaminated liquid and material is much less.

● The procedure is much faster.

● A horizontal chamber can be easily modified to serve as a TGGE apparatus (Suttorp *et al.* 1996).

In the following chapter washed and dried gels are rehydrated in urea-buffer. Gels can, of course, also be cast directly with urea and Tris-phosphate buffer as described in method 11.

1 Sample preparation

Microsatellite and AFLP samples

Denature samples by heating for 3 minutes at 95 °C with 50 % formamide (1 part PCR product + 1 part formamide) and place them in ice-water during the application procedure.

Differential display reverse transcription (DDRT):

For optimized amplification procedures work according to Bosch and Lohmann (1996). Separations are either performed in 15 %T short distance denaturing gels or in native long distance gels (see method 11).

Ribonuclease protection assay (RPA):

Prepare the samples with the help of commercial RNase protection assay kits and apply them directly on the gel.

Temperature gradient gel electrophoresis (TGGE):

The amplification products are pipetted directly on the gel.

2 Solutions

Dye: Prepare a 1 % (w/v) stock solution of bromophenol blue.

Gel buffer 0.36 mol/L Tris-phosphate pH 8.4 (4 × conc):

4.36 g Tris, dissolve in 80 mL H_2O_{dist}; titrate to pH 8.4 with phosphoric acid; make up to 100 mL with H_2O_{dist}.

Electrode buffer 450 mmol/L Tris / 75 mmol/L boric acid /
12.5 mmol/L EDTA-Na₂(5 × TBE):

54.5 g Tris + 23.1 g boric acid + 4.65 g EDTA Na₂,   make up
to 1 L with H₂O$_{dist}$.

Rehydration solution (90 mmol/L Tris-phosphate pH 8.3,
7 mol/L urea, 4 % glycerol, 4 % ethylenglycol):

10.5 g urea + 6.25 mL gel buffer + 1 mL glycerol + 1 mL
ethylenglycol + 0.5 mL bromophenol blue solution, make up to
25 mL with double-distilled water.

3 Rehydration

The preparation of washed and dried gels on carrier films is
described in method 11. For microsatellites, DDRT, and RPA
15% *T* gels are selected, for TGGE 10 % *T* short gels and for
AFLP 10 % *T* long gels are employed.

Lay GelPool onto a horizontal table; select the appropriate reswelling chamber, pipet the rehydration solution into the chamber, for

a complete gel:	25 mL
a half gel (short):	13 mL

Set the edge of the gel-film − with the gel surface facing
down − into the rehydration buffer (Fig. 2) and slowly lower
it, avoiding air bubbles.

The gels for short distances can be used in one piece, or − de-pending on the number of samples − cut into smaller porti-ons with scissors (when they are still dry). The rest of the gel should be sealed airtight in a plastic bag and stored in a free-zer.

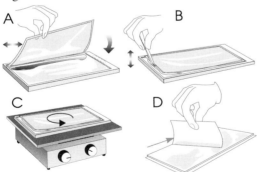

Fig. 2: Rehydration of a gel.
A Placing the dry gel into the GelPool. **B** Lifting the gel for an even distribution of the liquid.
C Rehydration on a rocking platform (not always necessary). **D** Removing the excess buffer
from the gel surface with filter paper.

Using foreceps, lift the film up to its middle, and lower it again
without catching air bubbles, in order to achieve an even distribution of the liquid (Fig. 2 B). Repeat this during the first 10 min.

When the gel surface is dry enough, this is indicated by a noise like a whistle.

90 min later 7 mol/ urea and the other additives have diffused into the gel, and the gel is removed from the GelPool. Dry sample wells with clean filter paper, wipe buffer off the gel surface with the edge of a filter paper (Fig. 2D).

4 Electrophoresis

Preparation of the electrode wicks

Short distance gels: Lay two of the 25 × 5 cm electrode wicks into the compartments of the PaperPool. Apply 20 mL of the respective electrode buffer to each wick (Fig. 3).

Do not forget, that for one buffer system different anode and cathode buffer is used.

Long distance gels: Cut six electrode strips of 11.7 ×1.8 cm. Lay three of them stacked into the compartments of the PaperPool. Apply 10.5 mL of the respective buffer to each stack.

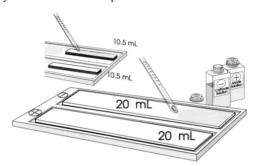

Fig. 3: Soaking the wicks with electrode buffer.

Application of the gel and the electrode wicks

Switch on the thermostatic circulator, adjusted to the 15 °C. Apply a very thin layer of kerosene (ca. 0.5 mL) onto the cooling plate with a tissue paper, in order to ensure good cooling contact (a few air bubbles do not matter).

Short distance run: Place the gel (surface up) on to the center of the cooling plate: the side containing the wells must be oriented towards the cathode (Fig. 4; Multiphor: line 5).

Place the cathodal strip onto the cathodal edge of the gel, edge of the electrode strip matching "3.5" on the cooling plate. The edge of the strip should be at least 4 mm away from the edges of the sample wells.

If the cathodal strip is place too close to the wells, small DNA fragments will exhibit less sharpness or hollow bands.

Place the anode strip over the anodal edge, matching "13.5" on the cooling plate. Smooth out air bubbles by sliding bent tip forceps along the edges of the wicks laying in contact with the gel.

Long distance run: Place the gel (surface up) on to the center of the cooling plate: the side containing the wells must be oriented towards the cathode . Place the cathodal stack electrode strips to the cathodal side of the gel, the inner edge matching with line "23" of the cooling plate.

Place the other stack over the anodal side, the inner edge matching with line "2" of the cooling plate. Smooth out air bubbles by sliding bent tip forceps along the edges of the wicks laying in contact with the gel.

Sample application and electrophoresis

Apply 6 µL of each sample to the sample wells using a micropipette. Clean platinum electrode wires before (and after) each electrophoresis run with a wet tissue paper. Move electrodes so that they will rest on the outer edges of the electrode wicks. Connect the cables of the electrodes to the apparatus and lower the electrode holder plate (Fig 4). Close the safety lid.

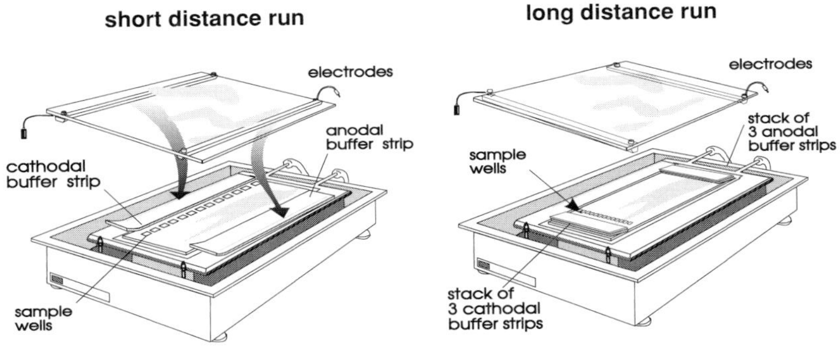

Fig. 4: Appliance for short and long distance runs of DNA fragments.

Running Conditions

The suggested running conditions have to be modified for some applications: e.g. when samples of the size range 700 to 900 bp have to be well separated, at least two hours separation time is required. The dyes added to the samples are a help for the estimation of the runnning time:

Gels, which contain 7 mol/L urea and run at 25 °C do not need to be covered with a film or a glass plate.

Short distance run in a whole15 % T gel,
Tris-phosphate / TBE buffer pH 8.4, 7 mol/L urea, 25 °C:

400 V_{max},	17 mA_{max}	6 W_{max},	15 min
1000 V_{max},	22 mA_{max}	25 W_{max},	1 h 20 min

Long distance run in a 10 % T gel,
Tris-phosphate /TBE buffer pH 8.4, 7 mol/L urea, 25 °C:

250 V_{max},	8 mA_{max}	5 W_{max},	30 min
1000 V_{max},	15 mA_{max}	12 W_{max},	1 h 50 min

Long distance runs:

Note: High voltage and relatively low milliampere values are   *see page 237*
applied on long electrophoresis gels. This means, that with a
milliampere-constant control only little changes in conductivity
can result in severe changes of the voltage value. In method 11
an alternative for an improved control for the running conditions
are described: voltage ramping.

Higher temperatures:

When higher temperatures like 40 to 60 °C have to be applied,
the gel surface has to be protected from drying. Suttorp *et al.*
(1996) cover it with a glass plate. In figure 5 an alternative with
cover films is shown.

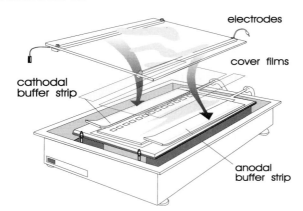

Fig. 5: Protection of the gel surface during high temperature runs with
cover films, which are applied after the samples have entered the gel.

5 Silver staining

It is suggested to employ the most sensitive silver stainnig
protocol as described on page 240 with one exception: The
fixation has to be modified for urea containing gels:
Fix: 30 min, 250 mL, 15 % ethanol + 5 % acetic acid (v/v)
at + 40 °C with the gel swimming on the surface of the liquid,
film-side up.

Method 15: Vertical PAGE

Polyacrylamide gel electrophoresis in a vertical setup is the standard technique in many laboratories. For the separation of proteins mostly discontinuous SDS electrophoresis according to Lämmli (1970) is performed, for DNA separations the continuous TBE (Tris borate EDTA) buffer is employed. Exact descriptions of gel casting and running conditions for gels of different sizes are found in the Hoefer Protein Electrophoresis Applications Guide (1994).

Hoefer Protein Electrophoresis Applications Guide (1994) 18-54.

Conventional procedure:

For a discontinuous system, the resolving gel is polymerized at least one day before use. After pouring the monomer solution into the cassette it is overlaid with water-saturated butanol to achieve a straight upper edge.

One hour before electrophoresis the butanol solution is removed, the edge is rinsed several times with a gel buffer solution to remove unpolymerized monomers, is dried with filter paper, the stacking gel solution is poured on top of the resolving gel, and the comb is inserted. After removal of the combs the wells are first rinsed and then filled using the upper buffer. The samples must contain at least 20 % v/v glycerol or sucrose and they are underlaid with a syringe or a fine-tipped pipette.

In the standard procedure the resolving and stacking gels are polymerized at different times.

In practice the interface between stacking and resolving gel frequently leads to problems: lateral edge effects, protein precipitate at the resolving gel edge, loss of th stacking gel during staining (probable loss of large proteins which can not migrate into the resolving gel).

Modified procedure:

Practice has shown, that the quality of results is not reduced, when the original procedures are modified as described below.

Reproducibility is even improved, when some steps are simplified.

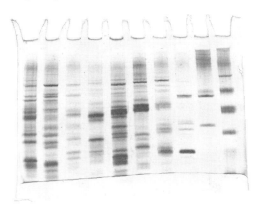

Fig. 1: Vertical SDS electrophoresis of legume seed extracts and markers, modified procedure; silver staining.

In the following chapter simplified procedures are described how to cast individual and multiple 0.75 mm thin gels – and to run them – for a vertical minigel system: Mighty Small chamber with 6 × 8 cm gels, see figure 1.

The principles described can be transferred to larger and thicker gels with no problems.

Instead of polymerizing the resolving gels and stacking gels separately, they are polymerized together. In order to prevent mixing of these monomer solutions during pouring them into the gel cassette, the resolving gel solution must contain a certain amount of glycerol. Glycerol in the monomer solution has no negative effects on the polymerization and the separation.

In this way, a lot of work can be saved: overlaying with butanol water, washing and drying the edge, casting the stacking gel before use.
Furthermore, no edge effects occur, the stacking gel does not fall off the resolving gel.

Gradient gels are cast without a pump with very good reproducibility.

This saves work and time.

1 Sample preparation

For SDS PAGE, the sample preparation is performed exactly as described in method 7 (page 167 ff) with the only difference, that – because of underlaying – the sample buffer must contain 25 % (v/v) glycerol:

For all vertical techniques, sample solutions must contain at least 20 % glycerol to prevent mixing with the cathode buffer.

Nonred SampB (Non reducing sample buffer):
1.0 g of SDS + 3 mg of EDTA + 10 mg of Bromophenol Blue + 2.5 mL of gel buffer*) + 25 mL glycerol (87 %), make up to 100 mL with distilled water.

*) In the original Lämmli procedure the stacking gel buffer is added; better results are achieved, when the resolving gel buffer is added here.

For native protein PAGE and DNA PAGE 25 % glycerol is just added to the samples.

2 Stock solutions

Acrylamide, Bis solution (T = 40%, C= 3%):
38.8 g of acrylamide + 1.2 g of Bis, make up to 100 mL with H_2O_{dist}. For vertical gels a 40 % T stock solution is used rather than a 30 % T solution (for the horizontal gels).

Because glycerol must be added to some highly concentrated starting solutions, higher concentrated acrylamide stock solutions are required.

Caution! *Acrylamide and Bis are toxic in the monomeric form.* Avoid skin contact and dispose of the remains ecologically.

Polymerize the remains with an excess of APS.

Stacking gel buffer pH 6.8 (4 × conc):
6.06 g of Tris + 0.4 g of SDS, make up to 80 mL with H_2O_{dist}. Titrate to pH 6.8 with 4 mol/L HCl; make up to 100 mL with H_2O_{dist}.

pH 6.8 for stacking gel is used only to achieve optimal polymerization conditions for samples wells (buffers will diffuse during storage).

Resolving gel buffer pH 8.8 (4 × conc):
18.18 g of Tris + 0.4 g of SDS, make up to 80 mL with H_2O_{dist}. Titrate to pH 8.8 with 4 mol/L HCl; make up to 100 mL with H_2O_{dist}.

Ammonium persulfate solution (APS):
Dissolve 400 mg of APS in 1 mL of H_2O_{dist}.

Can be stored for one week in the refrigerator (4 °C).

Push-up Solution:
11 mL Glycerol + 3.5 mL resolving gel buffer + 0.5 mL Orange
G solution (1 %).

Only needed for multiple gel casting.

Cathode buffer (10 × conc):
7.6 g of Tris + 36 g of glycine + 2.5 g of SDS, make up to
250 mL with H_2O_{dist}.

Do not titrate with HCl!

Anode buffer (10 × conc):
7.6 g of Tris + 200 mL of H_2O_{dist}. Titrate to pH = 8.4 with
4 mol/L HCl; make up to 250 mL with H_2O_{dist}.

Economy measure: the cathode buffer can also be used here.

3 Single gel casting

A gel cassette consists of a glass plate, a notched aluminum
oxide ceramics plate, two spacers and a comb (Fig. 2).

Aluminum oxide ceramics dissipates the heat much more efficiently than glass.

The gels must be prepared at least one day before use.

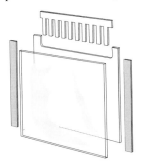

Fig. 2: Gel cassette for a vertical gel.

Important: In order to prevent drying of the stacking gels, the
gel cassettes should be taken out from the casting stand after one
hour of polymerization. They are placed into a plastic bag, to
which a few milliliters of gel buffer – diluted 1 : 4 with water –
are added.

The bags must be sealed and left at room temperature, when the gels are used next day, or in a refrigerator for longer storage.

When polyacrylamide gels contain a buffer with a pH value
above 7, the shelflives of these gels are limited to a couple of
weeks, because they hydrolize after some time and loose their
sieving properties. When gels are not needed every day, it is
recommended to cast only one or two gels, when needed.

Several weeks refrigerated; at room temperature maximum 10 days shelflife.

For the preparation of one or two gels a casting stand is used
with a rubber gasket bottom to seal the the gel cassette. The
cassettes are held together with two clamps (Fig. 3).

The comb is inserted after pouring the stacking gel solution.

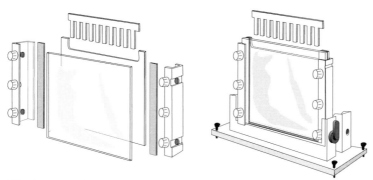

Fig. 3: Preparation of the gel casting stand for one or two gel cassettes.

Discontinuous SDS polyacrylamide gels

The difference of the pore sizes between stacking and resolving gel and the buffer ions are sufficient for a good stacking
effect.

see page 33 ff

Tab. 1: Composition of the gel solutions for two discontinuous gels

	Resolving gel 12 % T / 3 % C	Stacking gel 5 % T / 3 % C
Glycerol	2.0 mL	–
Acrylamide, Bis 40%T, 3%C	2.4 mL	0.5 mL
Res. Buffer	2.0 mL	–
Stack. Buffer	–	1.0 mL
TEMED	4 µL	2 µL
with H₂O_dist fill up	→ 8 mL	→ 4 mL
APS (40 %)	8 µL	4 µL

*This is an example for 0.75 mm
thin 12 % T gels and can easily
be recalculated for other gel
thicknesses and concentrations.*

First pipette 3.4 mL resolving gel solution into the cassette.
Then carefully apply 1.2 ml of stacking gel solution like an
overlay. Insert the comb without trapping air bubbles.

*Because of the difference in den-
sities of the solutions they do
not mix, a sharp interface is obtained.*

Porosity gradient gels

The casting procedure is similar to the technique used for
horizontal gradient gels: no pump is employed.

see page 174ff

The gradient is prepared with a *gradient maker* (s. Fig. 4). It
is made of two communicating cylinders.

The front cylinder, the *mixing chamber*, contains the denser solution and a magnetic stirrer bar. The back cylinder, the *reservoir*, contains the lighter solution. The dense solution contains about 25% glycerol and the light one 10%. The stacking gel solution – without glycerol – is overlaid over the gradient and copolymerized.

In the gradient presented here, the dense solution contains the higher proportion of acrylamide and gives the part of the gel with the smaller pores.

The *compensating bar* in the reservoir corrects for the difference in density and for the volume of the magnetic stirrer.

For reproducible gradients the outlet of the gradient maker must always be on the same level above the upper edge of the gel cassette.

The following recipe is an example for a gradient from $8\%T$ to $20\%T$ in 0.75 mm thin gels and can easily be adjusted to other gel thicknesses and concentrations.

Tab. 2: Composition of the gel solutions for two gradient gels.

Pipette into 3 test tubes	Dense solution $20\%T$ / $3\%C$	Light solution $8\%T$ / $3\%C$	Stacking gel $5\%T$ / $3\%C$
Glycerol	1.0 mL	0.5 mL	–
Acrylamide, Bis 40%T, 3%C	2.0 mL	0.8 mL	0.5 mL
Res. Buffer	1.0 mL	1.0 mL	–
Stack. Buffer	–	–	1.0 mL
TEMED	2 µL	2 µL	2 µL
with H2Odist fill up	→ 4 mL	→ 4 mL	→ 4 mL
APS (40 %)	4 mL	4 mL	4 mL

To pour a linear gradient (Fig. 4) both cylinders of the gradient maker are left open. The laboratory platform ("Laborboy") is set so that the outlet lies 5 cm above the upper edge of the gel. Before filling, the valve in the connecting channel between the *reservoir* and the *mixing chamber* as well as the *pinchcock* are shut. The stirring bar is then placed in the mixing chamber.

Note: The dense solution must contain less APS, in order to start the polymerization from the upper edge.

Thermal convection can distort the gradient, when polymerization starts from the bottom.

● Pour 1.7 mL of the light solution into the reservoir,

● briefly open the valve to fill the connecting channel,

● pour 1.7 mL of the dense solution into the mixing chamber,

● pipette APS into the reservoir, mix with compensating bar,

● pipette the APS into the mixing chamber and stir briefly but vigorously with the magnetic stirrer,

to disperse the catalyst

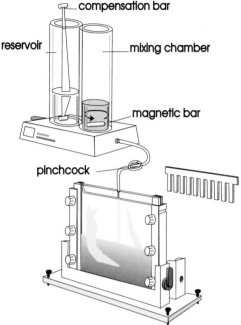

Fig. 4: Casting a gradient gel.

● place the tip of the tubing into the cassette,

● set the magnetic stirrer at moderate speed,   *do not generate air bubbles*

● open the connecting valve,

● open the outlet valve (pinchcock),

● when the gradient maker is empty, carefully apply 1.2 ml of stacking gel solution like an overlay,

● insert the comb without trapping air bubbles.

Rinse the gradient maker with distilled water immediately.

4 Multiple gel casting

The multiple casting stand is used with the silicon plugs inserted, thus there is almost no dead volume. First a plastic sheet is laid into the casting stand. The gel cassettes (see "2 Single gel casting") are placed into the casting stand – the ceramics plates to the back – with the combs already inserted (s. Fig. 5). It is strongly recommended to lay sheets of Parafilm® between the gel cassettes for their easy separation after the polymerization. After 12 gel cassettes have been inserted, two plastic sheets are added. The gasket is coated with a thin film of CelloSeal®. The cover plate is clamped to the stand.

The plastic sheets are used for filling the stand completely, and they make removing of the cassettes easier.

Parafilm® is much better than the wax paper.

When 1 or 1.5 mm gels are cast, only 10 respectively 8 gels can be prepared in one stand.

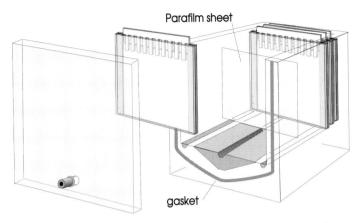

Fig. 5: Assembling the gel cassettes into a stand for multiple gel casting.

The solutions flow into the casting stand from below with a 35 cm long tubing connected to a gradient maker, also when no gradient is prepared. The laboratory platform ("Laborboy") is set so that the outlet lies 30 cm above the inlet of the casting stand. Before filling, the valve in the connecting channel between the reservoir and the mixing chamber as well as the pinchcock are shut.

Note: When the cassettes are packed in the correct way, the liquid flow can be observed through the front plate.

Multiple discontinuous SDS polyacrylamide gels

The difference of the pore sizes between stacking and resolving gel and the buffer ions are sufficient for a good stacking effect.

see page 33 ff

Tab. 3: Composition of the gel solutions for twelve discontinuous gels

	Resolving gel 12 % T / 3 % C	Stacking gel 5 % T / 3 % C
Glycerol	15 mL	–
Acrylamide, Bis 40%T, 3%C	18 mL	2.5 mL
Res. Buffer	15 mL	–
Stack. Buffer	–	5.0 mL
TEMED	30 µL	10 µL
with H₂O$_{dist}$ fill up	→ 60 mL	→ 20 mL
APS (40 %)	60 µL	20 µL

This is an example for twelve 0.75 mm thin 12 % T gels and can easily be recalculated for other gel thicknesses and concentrations.

Because of the difference in densities of the solutions they do not mix during casting; a sharp interface is obtained.

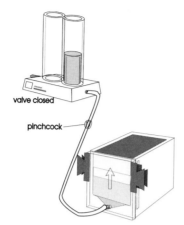

valve closed

pinchcock

Fig. 6: Casting multiple discontinuous gels.

● Pour 20 mL stacking gel solution into the cylinder of the gradient maker and let them flow into the stand.

● When the first air bubble leaves the outlet, immediately close the pinchcock.

● Fill 40 mL of the resolving gel solution into the cylinder and open the pinchcock again.

The liquid flows slowly because of the glycerol content.

● When ca. 1/2 half of the liquid has flown out, pour the rest into the cylinder.

● When the liquid level – of the stacking gel solution – has reached the edge of the ceramics plate, close the pinchcock.

● Empty the gradient maker by pouring the mixing chamber out into a beaker.

● Disconnect the tube from the outlet, connect it to a 1000 μL micropipette, open the pinchcock, press 1 mL of "Push-up solution" into the tube, close the pinchcock.

This measure keeps the tube clear of polymerization solution.

Rinse the gradient maker with distilled water immediately.

Do not forget to remove the gel cassettes from the casting stand after 1 hour of polymerization, and place them - with a few mL of gel buffer - into sealed plastic bags.

Otherwise the stacking gel starts to dry.

Multiple SDS polyacrylamide gradient gels

To pour a linear gradient both cylinders of the gradient maker are left open. As the solutions flow into the cassettes from below, the gradient maker is used with the light solution in the mixing chamber and the dense solution in the reservoir (see Fig. 7).

Note: The dense solution must contain less APS, in order to start the polymerization from the upper edge.

The compensation bar is not placed into the reservoir, because the stirrer bar compensates for the difference in densities.

Thermal convection can distort the gradient, when polymerization starts from the bottom.

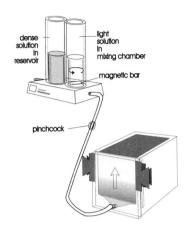

Fig. 7: Casting multiple gradient gels.

Tab. 4: Composition of the gel solutions for twelve gradient gels

	Dense solution 20 % T / 3 % C	Light solution 8 % T / 3 % C	Stacking gel 5 % T / 3 % C
Glycerol	6.25 mL	3.0 mL	–
Acrylamide, Bis 40%T, 3%C	12.5 mL	4.0 mL	2.5 mL
Res. Buffer	6.25 mL	6.25 mL	–
Stack. Buffer	–	–	5.0 mL
TEMED	25 µL	25 µL	10 µL
with H$_2$O$_{dist}$ fill up	→ 25 mL	→ 25 mL	→ 20 mL
APS (40 %)	20 µL	25 µL	20 µL

This is an example for twelve 0.75 mm thin gels with a gradient from 8 - 20 % T (see table 4); the solutions can easily be recalculated for other gel thicknesses and concentrations.

● Pour 20 mL stacking gel solution into the mixing chamber of the gradient maker and let them flow into the stand,

● when the first air bubble leaves the outlet, immediately close the pinchcock,

● place the stirring bar into the mixing chamber,

● pour the 25 mL dense solution into the reservoir,

● briefly open the valve to fill the connecting channel,

● pour the 25 mL light solution into the mixing chamber,

● pipette the APS into the reservoir and mix with the compensating bar,   *remove the compensating bar before casting*

● pipette the APS into the mixing chamber and stir briefly but vigorously with the magnetic stirrer,

to disperse the catalyst

● set the magnetic stirrer at moderate speed,

do not generate air bubbles

● open the connecting valve,

● open the outlet valve (pinchcock).

● When the gradient maker is empty, disconnect the tube from the outlet, connect it to a 1000 µL micropipette filled with 1 mL of "Push-up solution", open the pinchcock, press the solution into the tube, and close the pinchcock again.

This measure keeps the tube clear of polymerization solution.

Rinse the maker with distilled water immediately afterwards.

Do not forget to remove the gel cassettes from the casting stand after 1 hour of polymerization, and place them - with a few mL of gel buffer - into sealed plastic bags.

5 Electrophoresis

● Clamp the gel cassettes to the core with the notched ceramics plates facing to the center, the long side of the clamps on theb glass plate.

When a chamber for two gels is used and only one gel is run, clamp a glass plate or a ceramics plate to the opposite side of the core, to prevent a short circuit between the electrodes in the anodal buffer.

● Remove the combs.

● Mix 15 mL of cathode buffer (10×conc) with 135 mL deionized water and pour 75 mL into each cathodal compartment.

● Mix 15 mL of anode buffer (10×conc) with 135 mL deionized water and pour 150 mL into the anodal compartment.

● Load the samples using a pipette with standard tips: the tip is set on the edge of the ceramics plate and the sample is pushed slowly into the well (see Fig. 8).

● Place the safety lid on and connect to the power supply.

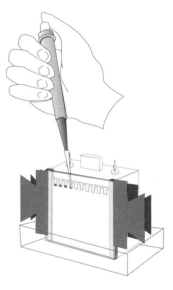

Fig. 8: Loading the samples on the vertical gel.

It is recommended to cool the gels with a thermostatic circulator: quicker separations are obtained, because higher field strength can be applied, and the results are better reproducible.

When the chamber is run in a cold room, cooling is much less effective.

Running conditions:

Direct cooling at 10 °C

Two gels 0.75 mm:	280 V_{max},	65 mA_{max},	18 W_{max}, 1 h
One gel 0.75 mm:	280 V_{max},	33 mA_{max},	9 W_{max}, 1 h

No direct cooling

Two gels 0.75 mm:	280 V_{max},	40 mA_{max},	12 W_{max}, 1 h 30 min
One gel 0.75 mm:	280 V_{max},	20 mA_{max},	6 W_{max}, 1 h 30 min

After the run:

● Switch off the power supply.

● Pour the anodal buffer out of the anodal compartment before opening the clamps, in order to avoid spilling of the buffer.

Open the clamps carefully and slowly remove the cassettes from the core.

● Open the cassettes only with a spacer or a plastic wedge.

A knife or a spatula damage the glass and ceramics plates.

● Clean and dry the glass and ceramics plates carefully.

6 SDS Electrophoresis of small peptides

The discontinuous gel and buffer system acc. to Schägger and Von Jagow (1987) provides a very good resolution of small peptides from 1 to 20 kDa. By using, a high gel buffer concentration (1 mol/L Tris), high crosslinking (6% C), pH 8.45 in both stacking and resolving gel, and replacing glycine in the cathodal buffer by tricine, an improved destacking of the small peptides is achieved.

As already mentioned in section I and in method 7, standard SDS PAGE shows a poor resolution of small peptides.

In the following part the recipe for buffers and the gels for a mini vertical system is described, which is derived from the method by Schägger and Von Jagow and the PhastGel® High Density; in the latter 30 % ethylenglycol is added to the polymerization solution.

With the ethylenglycol in the gel, the buffer concentration can be reduced to 0.75 mol/L, which results in a quicker separation.

Stock Solutions:

Anodal buffer 10×conc (2 mol/L Tris, HCl pH 8.9, 1 % SDS):
48.4 g Tris + 2.0 g SDS; make up to 160 mL with H_2O_{dist}; titrate to pH 8.9 with 4 mol/L HCl; fill up to 200 mL with H_2O_{dist}.

Note, that for the relatively low volume of anodal buffer a high Tris concetration is needed.

Cathodal buffer 10×conc (0.2 mol/L Tris,1.6 mol/L Tricine, 1% SDS):
3.87 g Tris + 56 g tricine + 2 g SDS; make up to 200 mL with H_2O_{dist}.

Gel buffer 4×conc (3.0 mol/L Tris / HCL pH 8.45):
36.3 g Tris + 0.4 g SDS; make up to 160 mL with H_2O_{dist}; titrate to pH 8.45 with 4 mol/L HCl; fill up to 100 mL with H_2O_{dist}.

Monomer solution (40 % T / 6 % C) for resolving gel:
37.6 g acrylamide + 2.4 g methylenbisacrylamide; fill up to 100 mL with H_2O_{dist}.

Store up to 3 months at room temperature in the dark.

Tab. 5: Composition of the gel solutions for two discontinuous gels for the analysis of small peptides.

	Resolving gel 16 % T / 6 % C	Stacking gel 5 % T / 3 % C
Ethylenglycol	2.8 mL	–
Acrylamide, Bis 40%T, 6%C	3.2 mL	–
Acrylamide, Bis 40%T, 3%C	–	0.5 mL
Gel buffer	2.0 mL	1.0 mL
TEMED	4 µL	2 µL
with H_2O_{dist} fill up	→ 8 mL	→ 4 mL
APS (40 %)	8 µL	4 µL

This is an example for 0.75 mm thin gels and can easily be re-calculated for other gel thick-nesses and sizes.

First pipette 3.4 mL resolving gel solution into the cassette. Then carefully apply 1.2 ml of stacking gel solution like an overlay. Insert the comb without trapping air bubbles.

Because of the difference in densities of the solutions they do not mix, a sharp interface is obtained.

Running conditions (10 °C):
200 Vmax   70 mA   18 W   2 hours 15 min

Figure 9 shows a typical separation result in a vertical peptide gel.

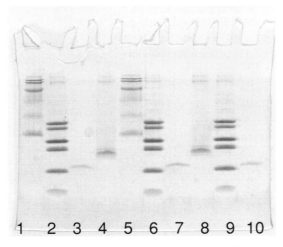

Fig. 9: SDS PAGE of small peptides. Samples: lanes 1,5 low molecular weight markers; lanes 2, 6, 9 peptide markers2.5 - 17 kDa; lanes 3, 7, 10 insulin; lanes 4, 8 aprotinin. Coomassie Brilliant blue staining.

7 Two-dimensional electrophoresis

The procedures of preparing the samples and IPG gels, and running the first dimension are thoroughly described in method 10.

see page 215ff

When the second dimension of an IPG-Dalt experiment should be run on a vertical gel, it is suggested to prepare either a disc gel with 12 %T or a gradient gel with 12 - 14 %T . Although larger gels are recommended for 2D electrophoresis, the principle is here described for minigels.

With the gradient no stacking gel is required.

The stacking or the gradient gel must be cast up to 0.5 cm below the edge of the ceramics plate:

This space is needed for embedding the IPG strip.

Multiple disc gels: Use only 10 mL stacking gel solution.
Single gradient gel: pipet 2.5 mL of each solution into the cylinders of the gradient maker.
Multiple gradient gels: pipet 35 mL of each solution into the cylinders of the gradient maker.

The recipes are easy to calculate or can be found in the Hoefer Application Guide (1994).

For a straight and well polymerized gel edge overlay each cassette with 300 µL of diluted gel buffer (1 : 4 with water). The equilibrated IPG strip is inserted into the cassette (see Fig. 10) and sealed in place with warm 0.5% agarose in sample buffer.

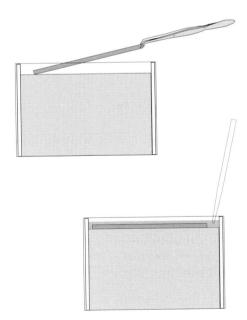

Fig. 10: IPG strip on the vertical second dimension slab gel.

8 DNA electrophoresis

For the separation of DNA fragments homogeneous polyacrylamide gels and mostly the continuous 90 mmol/L Tris-Borate-EDTA buffer are employed.

Tris-Borate-EDTA stock solution (4×conc):

0.36 mol/L Tris / 0.36 mol/L boric acid / 10 mmol/L EDTA-Na$_2$ (4 × TBE)

43.6 g Tris + 22.25 g boric acid + 3.72 g EDTA-Na$_2$; make up to 1L with distilled water.

For casting the gels the set-up described above is used, just without stacking gels. Because of the homogeneous buffer system, the separation is run at a constant voltage of 100 V.

These gels can be stained with silver staining or Ethidium bromide. *There are no fluorescent polyester films.*

9 Long shelflife gels

When a long shelflife is required, e.g. for multicast gels, the Tris-acetate / Tris -tricine buffer system can be employed:

Figure 1 on page 263 shows a separation, which has been performed with this buffer system.

Gel buffer 0.448 mol/L Tris-acetate pH 6.4 (4 × conc):
5.43 g Tris + 0.4 g SDS, dissolve in 80 mL H$_2$O$_{dist}$; titrate to pH 6.4 with acetic acid; make up to 100 mL with H$_2$O$_{dist}$.

With this buffer, the storage time of the gel is not limited. Because the pH value of the gel is lower than pH 7, the matrix does not hydrolyse.

Anode buffer (10 × conc):
7.58 g Tris + 0.25 g SDS, dissolve in 200 mL H$_2$O$_{dist}$; titrate to pH 8.4 with acetic acid; make up to 250 mL with H$_2$O$_{dist}$.

Cathode buffer (10 × conc):
4.84 g Tris + 71.7 g tricine + 0.25 g SDS, make up to 250 mL with H$_2$O$_{dist}$.

10 Coomassie and silver staining

When 0.75 mm thin gels are used, the procedures described in methods 7 and 10 for proteins and method 11 for DNA can be applied without any modifications.

see figure 1

For some low molecular weight peptides it is necessary to prevent diffusion out of or within the gel by an efficient fixing step prior to staining: 60 min in 0.2 % (w/v) glutardialdehyde + 0.2 mol/L sodium acetate in 30 % ethanol.

With this procedure the peptides are crosslinked in the polyacrylamide gel.

For *preservation* the gels are dried either on a filter paper with a gel dryer, which employes heat and vacuum, or between two sheets of wet cellophane, which are clamped in between two plastic frames.

A 1 Isoelectric focusing

1.1 PAGIEF with carrier ampholytes

Tab. A1-1: Gel characteristics.

Symptom	Cause	Remedy
Gel sticks to glass plate.	Glass plate too hydrophilic.	Clean the glass plate and coat with Repel Silane.
	Incomplete polymerization.	See below.
No gel or sticky gel, insufficient mechanical stability.	No or incomplete polymerization:	
	Poor water quality.	Always use double-distilled water!
	Too much oxygen in the gel solution (radical trap)	Degas thoroughly.
	Acrylamide, Bis or APS solutions too old.	Maximum storage time in the refrigerator: Acrylamide, Bis solution, 1 week; APS solution 40%, 1 week.
	Poor quality reagents.	Only use analytical grade quality reagents.
	Photochemical polymerization with riboflavin.	Chemical polymerization with APS is much more effective.
	The pH value is too basic (narrower basic pH range).	Rehydrate the prepolymerized and dried gel in a carrier ampholyte solution.
Gel peels away from the support film	Wrong support film was used.	Only use GelBond PAG film for polyacrylamide gels not GelBond film (for agarose).
	Wrong side of the support film was used.	Only cast the gel on the hydrophilic side of the support film, test with a drop of water.
	Support film was incorrectly stored or too old.	Always store the GelBond PAG film in a cool, dry, and dark place (< 25 °C), check the expiry date.

| | Insufficient polymerization. | See above. |
| | Gel solution contains non-ionic detergents (Triton X-100, Triton X-100). | Rehydrate the prepolymerized and dried gel in a carrier ampholyte/detergent solution (method 6) |

Tab. A1-2: Problems during IEF.

Symptom	Cause	Remedy
No current.	Safety turn off, "ground leakage" because of massive short circuit.	Turn off the power supply, check the separation unit and cable. Dry the bottom of the separation chamber, cooling coils and laboratory bench, turn the power on again.
Too low or no current.	Poor or no contact between the electrodes and electrode strips.	Make sure that the electrode strips are correctly placed; if a small gel or part of a gel is used, place it in the middle.
	The connecting cable is not plugged in.	Check the plug; press the plug more securely into the power supply.
Current rises during the IEF run.	Electrode strips or electrodes mixed up.	Acid solution at the anode, basic solution at the cathode.
General condensation.	The power setting is too high.	Check the power supply settings. Guide value: at most 1 W per mL of gel.
	Insufficient cooling.	Check temperature, if focusing is carried at a higher temperature e.g. 15 °C, reduce the power. Check the flow of the cooling fluid (bend in tubing?). Add kerosene between the cooling plate and the support film.
Condensation on the sample applicator.	Excessive salt concentration in the sample (>50 mmol/L) which causes local overheating.	Desalt the sample by gel filtration (NAP column) or dialyze against 1% glycin or 1% carrier ampholyte (w/v).
Gel swells around electrode strips.	Electroendosmosis causes a flow of water in direction of the electrodes (especially the cathode).	Normal phenomenon. It is not a problem unless it interferes with the run. It may help to occasionally blot the electrode strips.

	There is too much electrode solution in the electrode strips.	After soaking, blot the strips with filter paper.
	Electrode solution is too concentrated.	Use electrode solution at the specified concentration, dilute if necessary.
Condensation along the electrode strips.	Electrode strips reversed.	Acid solution at the anode, basic solution at the cathode.
Local condensation.	Localized hot spots due to bubbles in the insulating fluid.	Remove the air bubbles, avoid them from the beginning if possible.
Condensation over the basic half of the gel.	Electroosmotic water flow in direction of the cathode.	Use a better or fresher acrylamide solution. Reduce focusing time as much as possible, for narrow pH ranges pH > 7, blot the liquid which collects.
Lines of condensation over the whole gel.	Hot spots, conductivity gaps because of plateau phenomenon. Too long focusing time, especially for narrow pH ranges.	Fill the conductivity gaps by adding carrier ampholytes with a narrow pH range. Keep the focusing time as short as possible, or use IPG.
Sparking on the gel.	Same causes as for condensation, next stage (dried out gel).	Remedy as for condensation. Take measures as soon as condensation appears.
Sparking along the edge of the support film.	Electrode strips hang over the edge of the gel.	Cut the electrode strips to the size of the gel.
	High voltage and ions in the insulating fluid.	Use kerosene or DC-200 silicone oil, not water.

Tab. A1-3: Separations.

Symptom	Cause	Remedy
The pH gradient deviates from that expected.	Gradient drift (Plateau phenomenon).	
	Acrylic acid polymerized in the gel.	Only use analytical grade quality reagents.

	Acrylic acid polymerized in the gel because the acrylamide, Bis stock solutions were stored too long.	Maximum storage time in the refrigerator, in the dark: 1 week. The storage life can be prolongued by trapping the acrylic acid with Amberlite ion-exchanger MB-1.
	Temperature dependence of the pH gradient (pK values!).	Check the focusing temperature.
	Too long focusing time.	Reduce the focusing time as much as possible, especially in narrow basic pH intervals; or else use IPG.
	Gel stored too long.	Gels with narrow alkaline pH intervals have a limited storage life, use rehydratable gels.
	Gel contains carbonic acid ions.	Degas the rehydration solution (removal of CO_2); avoid the effects of CO_2 during IEF (particularly in basic pH ranges): seal the separation chamber, flush with N_2, trap CO_2: add 1 mol/L NaOH to the buffer tanks.
Partial loss of the most basic part of the pH gradient.	Oxidation of the carrier ampholytes during the run.	Reduce the influence of CO_2 as much as possible: see above.
	Oxidation of the electrode solutions.	See above.
Wavy iso-pH bands: 1. no influence of the sample.	Too much APS was used for polymerization.	Rehydrate a polymerized, washed and dried gel in a carrier ampholyte solution. Increase the viscosity of the gel by adding 10% (w/v) sorbitol to the rehydration solution or urea (< 4 mol/L, not denaturing in most cases).

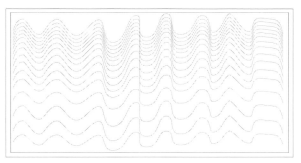

| | Urea gels stored too long, urea degraded to isocyanate. | Use urea gels immediately after preparation or else rehydrate the gel shortly before use. |

	Bad electrode contact.	Check the electrode contacts, especially the anode; if necessary put a weight on the electrode support.
	Unevenly or excessively wetted electrode strips.	Soak the electrode strips completely with electrode solution and blot them with filter paper.
	Wrong electrode solutions.	Use the electrode solutions recommended for the pH range in the correct concentrations.
	Gel too thin.	Ultrathin gels <200 μm are sensitive to protein overloading, varying protein concentrations, buffer and salt ions as well as diffusion of electrode solutions (compression of the gradient).
Wavy iso-pH lines. 2. induced by the sample.	Strongly varying protein concentration of the sample.	
(A) Protein concentration.		Either dilute the highly concentrated samples or apply in order of increasing or decreasing concentration. Prefocuse. Decrease the field strength at the beginning: (V/cm).
	Samples applied too far apart.	Apply samples closer to one another (1 – 2 mm)
	Highly concentrated samples applied at different places in the pH gradient.	It is not possible to proceed another way with the step trial test; but otherwise apply the samples as close as possible in the pH gradient. Prefocus; field strength at the beginning: < 40 V/cm.
	Overloading, – protein concentration too high.	Dilute the samples (2 to 15 μg per band) or use a thicker gel. Ultrathin-layer IEF: use gel thickness >250 μm or IPG.
(B) Buffer, salt concentration.	The buffer or salt concentrations in the samples vary a lot.	As for (A); if necessary, desalt highly concentrated samples.

	High buffer or salt concentration in the samples.	As for (A); make sure that the samples are applied close to one another, at the same level in the gradient. Prefocus; field strength at the beginning < 40 V/cm; at first for 30 min at < 20 V/cm; let salt ions migrate out of the gel or use IPG.
	Buffer or salt concentrations in the samples too high - desalting too risky or not possible because of eventual protein losses.	Cast sample application strips in polyacrylamide (T = 10% ca.) or agarose (ca. 2%) containing salts in the same concentrations as the samples and place over the whole width of the gel. Apply the samples in the wells; let the salt ions migrate out for about 30 min at E = V/cm, the salt load will then be the same over the whole width of the gel, so individual shifts in the pH gradient will be compensated. Alternatively use IPG.
Streaking or tailing of the sample.	Precipitate and/or particles in the sample.	Centrifuge the sample.
	The applicators retain the proteins and release them later.	Remove the applicators after about 30 min of IEF or use applicator strips.
	Old or denatured sample.	Check the sample preparation procedure, carry it out shortly before the separation. Store samples at <−20 °C.
	High molecular weight proteins have not reached their pI yet.	Focus longer or use agarose gels.
	Poorly soluble proteins in the sample.	Focus in urea (if necessary with non-ionic or zwitterionic detergents) or 30% DMSO.
	Protein overloading.	Dilute sample or apply less.
	Protein aggregation during sample entry.	Set a lower current (mA) limit for the sample entry phase. This reduces the field strength in the beginning.

Diffuse bands.	Diffusion during IEF, low molecular weight peptides.	It is preferable to focus oligopeptides with molecular weight < 2 kDa in IPG where diffusion is less marked.
	Diffusion after IEF, inadequate or reversible fixing.	Check the fixing and staining methods.
	Urea IEF: Urea precipitation in the gel.	Run urea gels at 15 - 20 °C.
	Focusing time too short.	Focus for a longer time.
	Marked gradient drift.	See above.
	Influence of CO_2 on the basic bands.	See above.
Individual bands are diffuse.	See above.	See above.
	The focusing time for the individual proteins is too short (large molecules and/or low net charge).	Optimize the sample application point with a concentration test or titration curve analysis. Apply the sample on the side of the pI where the charge curve is steeper.
Missing bands.	Concentration too low or detection method not sensitive enough.	Apply more sample or concentrate the sample. Use another detection method (e.g. silver staining of the dried gel, blotting).
	The proteins are absorbed on the sample applicator.	Use sample application strips.
The proteins precipitate at the point of application.	Application too close to the pI.	Apply sample further away from its pI (step trial test, titration curve).
	The field strength is too high at the point of sample entry.	Reduce the voltage at the beginning (E< 40 V/cm).
	The molecule is too large for the pores of the gel.	Use agarose instead of polyacrylamide.

	The proteins form complexes.	Add urea (7 mol/L) to the sample and the gel; add EDTA to the sample; add non-ionic or zwitter-ionic detergents to the sample and the gel.
	Protein unstable at the pH of site of application.	Apply the sample at another point (step trial test, titration curve).
	The protein is unstable at the temperature used.	Change the focusing temperature.
Individual bands focus at the wrong place.	The proteins form complexes.	See above. If it is suspected that complexes form with the carrier ampholyte, check with IPG.
	The proteins have lost ligands.	Check with titration curve analysis.
"One" protein focuses in several bands.	The protein exists in various states of oxidation.	Check the sample preparation; eventually focus under N_2.
	The protein has dissociated into subunits.	Do not focus in the presence of urea.
	Urea IEF: Carbamylation by cyanate.	Check sample preparation and gel casting with urea.
	Different conformations of a molecule.	Focuse the protein in the presence of urea (>7 mol/L).
	Different combinations of oligomers of a protein or of subunits.	Natural phenomenon.
	Different degrees of enzymatic phosphorylation, methylation or acetylation exist.	Check the sample preparation procedure.
	Various carbohydrate moieties of glycoproteins.	Natural phenomenon. Treat the sample with neuramidase for example, to verify.
	Partial proteolytic digestion of a protein.	Check the sample preparation procedure. Add inhibitor (e.g. 8 mmol/L PMSF).
	Complex formation.	If complex formation with the carrier ampholytes is suspected, verify with immobilized pH gradients.

1.2 Agarose IEF with carrier ampholytes

Only problems specific to agarose IEF are mentioned here. Consult paragraph 1.1 under PA-GIEF for general problems occurring in connection with isoelectric focusing or carrier ampholytes.

When agarose IEF is used, it should be remembered that the matrix is not as electrically inert as polyacrylamide since sulfate and carboxylic groups are still bound to agarose which is of natural origin, and they give rise to electroendosmotic phenomena.

Tab. A1-4: Gel properties.

Symptom	Cause	Remedy
Insufficient gel consistency.	Incomplete solidification of the gel.	Let the gel solidify > 1 h. It is best to remove it from the cassette after 1 h and store it overnight in a humidity chamber at +4 °C (maximum storage time: 1 week).
	The agarose concentration is too low despite the fact that the agarose was precisely weighed out, the agarose has absorbed water.	Store the agarose in a dry place out of the refrigerator. Close the package well.
	Urea gel: urea disrupts the structure of agarose	Use a higher agarose concentration (2%); let the gel solidify longer or use rehydratable agarose gels.
The gel comes off the support film.	Wrong support film used.	Only use GelBond film for agarose, not GelBond PAG film (for polyacrylamide gels).
	The wrong side of the support film was used.	Cast the gel on the hydrophilic side of the support film.
	The gel was cast at too high a temperature. The solidification time was too short.	The temperature should be kept between 60 and 70 °C during casting. See above.

Tab. A1-5: Problems during the IEF run.

Symptom	Cause	Remedy
Flooding on the surface.	The gel surface was not dried. The solidification time was too short.	Always dry the surface of the gel with filter paper before IEF. See above.

	The wrong electrode solution was used.	In general it is recommended to use: at the anode: 0.25 mol/L acetic acid; at the cathode 0.25 mol/L NaOH.
	The electrode strips are too wet.	Remove the excess liquid. Blot the electrode strips so that they appear almost dry.
	Electroendosmosis.	Natural phenomenon. Blot the electrode strips every 30 min. or replace them by new ones.
	Marked electroendosmosis.	Always use double-distilled water; use an ideal combination of chemicals: 0.8% agarose IEF with 2.7% Ampholine.
	There are no water binding additives in the gel.	Add 10% sorbitol to the gel solution.
Water build-up at the cathode.	Electroendosmosis, cathodic drift.	Dry the cathode strips more often and carefully; only focus as long as necessary.
Water build-up at the sample application site.	Electroendosmosis because of the material used for sample application.	Only use sample application strips or masks, do not use paper or Paratex for example.
	The protein or salt concentration is too high.	See polyacrylamide gels.
Formation of a ditch in the gel.	Advanced cathodic drift because of electroendosmosis.	See above. Focus at 10 to 15 °C.
	Insufficient gel consistency.	See above.
Formation of small hollows near the sample application site.	The power was too high during sample entrance.	Set the power at 5 to 10 W at most for the first 10 to 15 min (for a 1 mm thick gel, 25 cm wide × 10 cm separation distance, use correspondingly lower settings for smaller gels.
	Sample overloading.	See under polyacrylamide gels.
The gel dries out.	Advanced electroendosmosis.	See above.
	The gel was irregularly cast.	Position the leveling table exactly when casting horizontal gels or else use the vertical technique ("clamp" technique in prewarmed molds).

	Heat source in the proximity.	During agarose IEF do not place the separation chamber beside a thermostatic water bath.
	The air is too dry.	When the ambient humidity is too low, pour a small volume of water in the electrode tanks.
Sparking.	Advanced stage of the effects listed above.	See above, if possible take measures before this occurs.

Tab. A1-6: Separation results.

Symptom	Cause	Remedy
Bands too wide.	Too much sample solution was applied.	Reduce the sample volume.
Diffuse bands.	Focusing time too long (gradient drift) or too short (the proteins have not reached their pI yet).	See above and under polyacrylamide gel.
	Because of the larger pore size, diffusion is more marked in agarose than in polyacrylamide gels.	Check the fixing and staining procedures. Dry the gel after fixing and then stain (this is also valid for silver staining).
Missing bands.	See above.	See above.
Missing bands in the basic part of the gel.	Part of the gradient is lost because of a cathodic drift (more pronounced in agarose than in polyacrylamide gels).	Add a carrier ampholyte with a narrow basic range; focus for a shorter time.
Distorted bands at the edge of the gel.	Fluid has run out of the gel or the electrode strips; fluid has run along the edge of the gel and forms L-shaped "electrodes".	Blot the gel or electrode strips regularly when water oozes out.
	The samples were applied too close to the edge.	Apply the samples about 1 cm from the edge.
Wavy bands in the gel.	As for polyacrylamide gels.	See under polyacrylamide gels.
	Irregularities in the surface of the gel.	Degas the gel solution properly. Use a humidity chamber for storage.

Tab. A1-7: Problems specific to agarose IEF.

Symptom	Cause	Remedy
Bands are diffuse, disappear or do not appear.	Diffusion.	Always dry agarose gels after fixing and before staining them.
The gel comes off the support film during staining.	The fixing solution was not completely removed from the gel before drying. A mistake was made during gel casting.	Rinse the gel in twice for 20 min each time in the destaining solution containing 5% glycerol. See above.

1.3. Immobilized pH gradients

Tab. A1-8: Gel properties.

Symptom	Cause	Remedy
The gel sticks to the glass plate.	The glass plate is too hydrophilic.	Clean the glass plate and coat it with Repel Silane.
	The gel was left too long in the mold.	Remove it from the cassette 1 h after the beginning of polymerization.
	The gel concentration is too low. Incomplete polymerization.	Do not use glass when T < 4%, use acrylic glass (Plexiglas) instead. See below.
No gel or sticky gel.	Poor water quality. The acrylamide, Bis or APS solutions are too old. Poor quality reagents.	Always use double-distilled water! Maximum storage time in the dark in the refrigerator: acrylamide, Bis solution: 1 week. 40% APS solution: 1 week. Only use reagents of analytical grade quality.
	Too little APS and/or too little TEMED were used.	Always use 1 µL of APS solution (40% w/v) per mL of gel solution and at least 0.5 µL of TEMED (100%) per mL of gel solution.
	The pH value was not optimal for polymerization.	For wide (1 pH unit) and alkaline (above pH 7.5) pH ranges: titrate both gel solutions with HCl 4 mol/L respectively NaOH 4 mol/L (or 100% TEMED) to about pH 7 after TEMED has been added. The precision of pH paper is sufficient.

	The polymerization temperature was too low.	Let the gel polymerize for 1 h in a heating cabinet or incubator at 50 °C or 37 °C respectively.
One half of the gel is not or insufficiently polymerized.	The APS solution has not mixed properly with the gel solution (usually the dense solution: the APS solution overlayers it because of the glycerol content).	After adding the APS solution stir vigorously for a short time. Make sure that the drops of APS solution are incorporated in the gel solution.
	One of the solutions was not titrated to pH 7.	See above.
The surface of the gel is sticky, swells during washing and detaches itself from the support film.	Oxygen has inhibited polymerization of the surface.	Overlay the surface of the gel with about 300 μL of double distilled water immediately after casting; do not use butanol.
The gel detaches itself from the support film.	The wrong support film or the wrong side of the support film were used or else the support film was stored incorrectly.	See under polyacrylamide gels.

Tab. A1-9: Effects during washing.

Symptom	Cause	Remedy
The gel has a "snake skin" structure in certain areas or all over.	This is normal. Because of the fixed buffering groups the gel possesses slight ion-exchanger properties and swells.	The gel surface will become normal again when it is dry.
The gel becomes wedge shaped.	This is normal. The buffer has different concentrations and properties which results in different swelling characteristics within the gradient.	Dry the gel after washing; rehydrate it in the reswelling cassette (the cassette prevents it from taking a wedge shape).

Tab. A1-10: Effects during drying.

Symptom	Cause	Remedy
The support film rolls up.	The gel pulls in all directions.	Add 1% to 2% of glycerol to the last washing, this makes the gel more elastic.

Tab. A1-11: Effects during rehydration.

Symptom	Cause	Remedy
The gel does not swell or only partially.	The reswelling time is too short.	Adapt the reswelling time to the additive and the additive concentration. If the gel was stored for a long time at room temperature or if the use-by date is expired, prolong the reswelling time.
	Gel was dried too long or at too high a temperature.	Dry the gel with a fan at room temperature, the air-flow should be parallel to the gel surface and the gel should be dried in a dust free atmosphere
	The gel was stored too long at room temperature or higher.	Use the gel immediately after drying or store it hermetically sealed at < -20 °C.
The gel sticks to the reswelling cassette.	The surface of the glass is too hydrophilic.	Coat the surface of the gel within the gasket with Repel Silane.
The gel sticks to the support glass plate.	The gel surface was inadvertently applied on the wet glass plate.	Pull it away gently under water in a basin.

Tab. A1-12: Effects during the IEF run.

Symptom	Cause	Remedy
No current.	The cable is not plugged in.	Check the plug; insert the plug more securely in the power supply.
Low current.	This is normal for IPG. The gels have a very low conductivity.	Standard setting for whole IPG gels: 3500 V, 1.0 mA, 5.0 W. Regulate IPG strips with the voltage setting.
Localized condensation over specific areas.	Salt concentration in the samples is too high. Salt ions form arcs when leaving the sample wells, spots with very high salt concentration result where two fronts meet.	Apply samples with high salt concentration close to one another, if samples must be applied at different areas within the pH gradient. Separate the traces by cutting strips or scraping out troughs.

Local sparking at specific points.	See above; next stage.	See above; if focusing is carried out overnight, do not apply more than 2500 V and turn up to 3500 V the next day.
Sparking along the edge of the gel.	High voltage and there are ions in the contact fluid.	Use kerosene as contact fluid between the cooling plate and the film.
Sparking at an electrode.	The gel has dried out because of electroendosmosis. This occurs in narrow pH gradient s at extreme pH intervals (< 4.5; > pH 9).	Either add glycerol (25%) or 0.5% non-ionic detergent to the reswelling solution.
	The electrode solutions are too concentrated.	Soak both electrode strips in double-distilled water. The conductivity is sufficient; in addition, the field strength decreases at the beginning of IEF for improved sample entrance .
	Gel insufficiently polymerized.	See above.
A narrow ridge develops over the whole width of the gel and slowly migrates in direction of an electrode.	This is a normal phenomenon during IPG: it is an ion front at which a jump in the ionic strength and a reversal of the electroendosmotic effect occur.	Wash the gel thoroughly. Apply the sample so that the front comes from the furthest electrode. Add 2 mmol/L acetic acid to the reswelling solution for samples applied at the anode and 2 mmol/L Tris to the samples applied at the cathode.
The ridge does not migrate any further.	The gel contains too many free ions, the difference in conductivity within the gel is so large that the voltage is not sufficient to carry the ions further.	Wash the gel thoroughly. Use a power supply with a high voltage (3500 V are sufficient). Focus for a long time, overnight if necessary.

Tab. A1-13: Separation results.

Symptom	Cause	Remedy
The bands and iso-pH lines form arcs in the gel.	The gel polymerized before the concentration gradient had finished leveling.	Cool the casting cassette in the refrigerator before casting (this delays the onset of the polymerization). Use glycerol and not sucrose (its viscosity is too high) to make the acid solution denser.

	The catalyst was not properly washed out.	See above.
The bands are diffuse.	Focusing time too short.	Focus for a longer time, overnight for example.
	The field strength is not sufficient when the pH range is narrow or the separation distance is long (10 cm).	High voltages are necessary for narrow pH ranges and long separation distances: use a 3500 V power supply.
	There are problems with polymerization, for example the acrylamide or Bis solutions are old; see above.	Use fresh stock solutions.
The bands in the basic part of the gel are diffuse.	Influence of CO_2.	Trap CO_2 during IEF: seal the chamber, add soda lime or 1 mol/L NaOH to the buffer tanks.
No bands are visible.	The pH gradient is wrongly orientated.	Place the gel on the cooling plate with the acid side towards the anode and the basic side towards the cathode; the basic side has an irregular edge and the support film sticks out.
The proteins have stayed at the site of application.	The field strength was too high at first.	Do not prefocus (the pH gradient already exists). Keep the field strength low at the beginning.
	The proteins have aggregated at the site of application because their concentration was too high.	Dilute the sample with water or water/non-ionic detergent; it is preferable to apply a large sample volume than a concentrated solution.
	Some proteins have formed complexes and obstructed the pores.	Add EDTA to the sample. Add urea to the sample and rehydration solution; complex formation is prevented by a urea concentration of 4 mol/L but most enzymes are not denatured yet.
	Conductivity problems.	Apply the sample to the other side, or direct the ionic front as described above.
	The salt concentration in the sample is too high.	Dilute the sample with water and apply a larger sample volume.

	High molecular weight proteins are unstable when the ionic strength is low.	Prepare a gel matrix with large pores so that the protein can penetrate the gel before it has completely separated from the low molecular substances. As emergency measure it is recommended to add 0.8% (w/v) carrier ampholyte to the sample and 0.5% (w/v) carrier ampholyte from the corresponding pH range to the rehydration solution.
The pI of the proteins lies outside of the immobilized pH gradient.	Narrow pH range: the focusing was carried out at the wrong temperature.	Focus at 10 °C and/or widen the pH range.
	The pI obtained by carrier ampholyte IEF is shifted in comparison to the one obtained by IPG.	Use a wider or different pH range.
The immobilized pH range is not correct or not present.	Immobiline was not stored correctly. The acrylamide or Bis solutions are too old. A mistake was made when the Immobiline was pipetted.	Follow the recipes for Immobiline and casting instructions exactly; otherwise: see above.
	The focusing time is not sufficient.	Lengthen the focusing time, if necessary focus overnight.
Some bands are missing, are diffuse or are at the wrong place.	Oxygen sensitive proteins have oxidized in the gel (Immobiline gels trap oxygen from the air during drying).	Add a reducing agent to the rehydration solution when working with proteins which are sensitive to oxygen.
The separation lanes are curved and run from one another.	The conductivity of the gel is much lower than the conductivity of the sample (proteins, buffer, salts).	Direct the ionic front as described above; apply the samples beside one another; separate the lanes by cutting the gel or scraping out troughs.

Table A1-14: Specific staining problems during IPG.

Symptom	Cause	Remedy
There is a blue background after Coomassie staining.	Basic Immobiline groups tend to bind Coomassie.	Use a solution with 0.5% Coomassie; or, even better, use colloidal staining: no background staining!

A2 SDS-electrophoresis

Tab. A2-1: Gel casting.

Symptom	Cause	Remedy
Incomplete polymeriza-tion.	Poor quality chemicals.	Only use analytical grade quality rea-gents.
	The acrylamide and/or APS solutions were kept too long.	Always store the stock solutions in the refrigerator in the dark; the 40% APS solution can be stored for one week, solutions of lower concentrati-on should be freshly prepared every day; in case of doubt make new stock solutions.
	The water is of poor quality.	Always use double-distilled water.
The gel sticks to the glass plate. Silane.	The glass plate is too hydrophilic.	Wash the glass plate and coat it with Repel.
Leakage from the gra-dient mixer .	The rubber gasket is dry.	Open the gradient mixer, coat the gas-ket with a thin layer of CelloSeal®.
Gradient gels: the gel solution already poly-merizes in the gradient mixer.	Too much APS was used.	Reduce the amount of APS. Open the gradient mixer and clean it.
Air bubbles are trapped in the cassette.	This cannot always be prevented.	Carefully pull them out with a strip of film.
Gradient gels: one half of the gel is not or is in-completely polymerized.	The APS solution has not mixed with the gel solution (it has stayed on the sides or on the dense solution).	Pipette carefully; stir vigorously for a short time so that the APS solution is drawn into the dense solution.
The gel separates from the support film.	Wrong support film was used.	Only use GelBond PAG film for po-lyacrylamide gels not GelBond film (for agarose).
	The wrong side of the support film was used.	Cast the gel on the hydrophilic side; test with a drop of water.
	The support film was wrongly stored or too old.	Always store the GelBond PAG film in a cool, dry and dark place.

There is a liquid film on the surface.	The polyacrylamide matrix has hydrolysed because the buffer is too alkaline (pH 8.8).	Alkaline polyacrylamide gels should not be stored longer than 10 days in the refrigerator.
Holes in the sample wells.	Air bubbles were incorporated during casting.	Cut out the slot former with a sharp scalpel. Press the cut edge down with a pair of curved tweezers; use "crystal clear" Tesa film or Dymo tape with a smooth surface; otherwise small air bubbles which inhibit polymerization in their vicinity can form.
Holes in the gel ("Swiss cheese effect").	Many very small bubbles in the gel solution.	Degas the gel solution; when casting gradient gels do not stir too fast, because SDS solutions foams.
The edges of the gel have not polymerized enough.	The gel solution was not overlayed.	Overlay the gel solution.
	Oxygen from the air has diffused through the seal.	Polymerize the gel at a higher temperature (37 to 50 °C, ca. 30 min).
	Polymerization is too slow.	Degas the gel solution; add a little more TEMED and APS.

Tab. A2-2: Effects during electrophoresis.

Symptom	Cause	Remedy
Droplets on the surface of the gel.	Buffer dripped on the surface of the gel when the electrode wicks soaked in buffer were placed on the surface of the gel.	Make sure that the electrode strips are never held over the surface of the gel; if this happens, carefully remove the drops with a bit of filter paper.
	High ambient temperature and high humidity, water condensation on the surface of the cooled gel. (summer time !).	Apply gel and samples on the cooling plate before the cooling waterbath is connected to the cooling plate. Do not connect to the cooling system before electrodes and safety lid cover the gel.
Sample applicator strip: the samples merge together.	The sample applicator strip is not applied well enough.	Press the sample applicator down properly, do not touch it anymore, do not touch the strip with the tip of the pipette when applying the sample; use sample application pieces.

Sample wells: the samples leave the wells and spread over the gel surface.	Glycerol oder sucrose in the sample, osmotic distribution.	Prepare samples without glycerol or sucrose; those are only necessary for vertical PAGE.
	High protein content, sample contains proteins with low surface tension.	Add 8 mol/L urea to each sample (**after** heating!), the urea does not influence the electrophoresis pattern.
No current.	The electrode cable is not plugged in.	Check whether all the cables are properly connected.
The power supply switches itself off and exhibits ground leakage.	Electricity is leaking from the chamber.	Make sure that the laboratory bench is dry; if the ambient temperature and humidity are high, regularly wipe off the condensation water from the tubings: ideally use a foam rubber tubing to cover the cooling tubing.
The current decreases quickly, the voltage increases quickly.	The system runs out of buffer ions, because the electrodes are placed too close together.	Place electrodes as far as possible to the outer edges of the buffer strips, in order to include the complete buffer between the electrodes.

The front migrates in the shortest side of the gel.	The plugs are inverted: the gel is wrongly oriented.	Check whether the cable is plugged in properly; place the gel so that the sample application point lies near the cathode.
The front migrates too slowly; the separation takes too long.	The current flows under the support film.	Use kerosene or DC-200 silicone oil as contact fluid, not water.
The electrophoresis takes too long.	There are chloride ions in the cathode buffer.	The cathode buffer (Tris glycine) must not be titrated with HCl even if the pH value given in the recipe (usually pH 8.3) is not reached with glycine; the pH usually sets itself at 8.9 which is all right.
Condensation.	The power is too high.	Check the setting of the power supply: at most 2.5 W/mL of gel.

	The cooling is insufficient.	Check the cooling temperature (10 to 15 °C are recommended); check the flow of the cooling fluid (bend in the tubing?); add contact fluid (kerosene) between the cooling plate and the support film.
The front is crooked.	Irregular electrical contact.	See above.
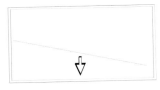	The buffer concentration in the electrode strips is irregular because they were not held straight when they were soaked or placed on the gel.	Make sure that the electrode wicks are always held straight.
	The casting mold was not levelled when the disk or gradient gels were cast.	Set the mold level with a spirit-level.
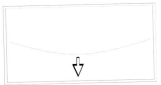 The front is curved.	The gel polymerized before the density gradient had settled or before the density discontinuity was finished.	Delay the onset of the polymerization: either by reducing the quantity of APS and/or by cooling the casting cassette in the refrigerator.
The front is uneven or wavy.	The gel surface has dried out in places under the holes in the lid.	Place a glass plate or electrode holder over the electrode wicks and the gel.
Formation of white precipitates and an irregular gel surface.	There is dirt in the electrode wicks which forms a precipitate with SDS.	Use electrode paper of the best quality; only handle the paper with rubber gloves.

The gel dries along one edge of the paper and burns through there.	This is an electroendosmotic effect due to poor quality of the chemicals and/or an old acrylamide solution.	Only use chemicals of analytical grade quality; store the acrylamide stock solution in the refrigerator in the dark for a short time (at most 2 weeks for SDS gels).
The slots dry out and burn through after a while; at the same time water collects at the cathodic side of the slots.	The samples were wrongly prepared; free SH groups led to the formation of disulfide bridges between the various polypeptides; these aggregates are too large for the gel pores and highly negatively charged (SDS), → electroosmotic flow of water towards the cathode.	Protect DTT from oxidation with EDTA; after reducing, heating and recooling the samples, add the same quantity of DTT as for the reduction; or else alkylate with iodoacetamide.

Tab. A2-3: Separation results.

Symptom	Cause	Remedy
The bands are not straight but curved.	Part of the gradient was shifted by heat convection during polymerization.	Add a little more APS to the light solution than to the dense one so that the polymerization starts at the top and proceeds towards the bottom.
The bands are close together at the buffer front.	The concentration of the resolving gel is too low.	Increase the concentration of the resolving gel.
	Low molecular weight peptides are poorly resolved.	Use gradient gels (if necessary concave exponential pore gradients); or use the buffer system of Schägger and von Jagow (1987).
The low molecular weight proteins are missing.	The post-polymerization is not finished.	Polymerize the gel at least one day ahead.
The bands are "fuzzy" and diffuse.	The post-polymerization is not finished.	Polymerize the resolving gel at least one day ahead.
	The resolving gel is not properly polymerized.	Always prepare the resolving gel at least one day before use, because a slow post-polymerization takes place in the gel matrix.

The bands are diffuse.	The proteins were applied too close to the cathode.	Apply the samples at least 1 cm from the edge of the cathode wick or the buffer strips.
	The samples are old.	Prepare fresh samples; it is helpful to boil the samples again (adding reducing agent before and after); or else alkylate them.
	A homogeneous gel system was used or else the buffer system was not optimal for the samples.	Try disc PAGE and/or gradient gel PAGE or another buffer system.
	Insufficient gel polymerization.	See above.
Artificial double bands.	Partial refolding of the molecules because the SH groups are not sufficiently protected.	Alkylate the samples after reduction.
The sample concentration is irregularly distributed in the bands.	Very small sample volumes were applied so the filling of the sample wells was irregular.	Dilute the samples with buffer and apply a correspondingly larger volume or apply the sample directly on the surface of the gel (for < 5 µL).
Crooked bands.	There is too much salt in the samples which leads to marked conductivity differences at the beginning.	Desalt the samples by gel filtration (NAP column) or by dialyzing against sample buffer.
Precipitates at the edge of the slots.	Gradient gel: the gradient was poured the wrong way round and the samples applied in the part with narrow pores.	Make sure that the slots are in the area with the lower acrylamide concentration.
	Electro-decantation, the slots are too deep.	Make the slots less deep (decrease the number of layers of Tesa film), they should not be more than 1/3 of the thickness of the gel. If larger sample volumes must be applied, it is preferable to increase the surface of the slots.
	A few proteins are too large to penetrate the gel pores.	Use gradient gels which allow separate of a broader molecular weight spectrum.

	Mistake during sample application; the free SH groups were not protected and disulfide bridges have formed between the polypeptides which are now too large for the pores.	Protect DTT from oxidation with EDTA; after reduction, heating and cooling, add the same amount of DTT as for the reduction to the samples; or else alkylate them, with iodoacetamide for example.
High molecular weight "ghost bands".	Mistake during sample preparation, see above.	See above.
Protein precipitate at the edge of the sample well; a narrowing then widening trace stretches below the sample well through the separation trace.	Overloading; the polypeptides form complexes and aggregate when entering the gel matrix; protein dissolves intermittently and migrates towards the anode.	Dilute the sample, apply less.
The bands streak.	There are particles in the sample.	Centrifuge or filter the sample.
There are individual streaks which sometimes begin in the middle of the gel.	Dust particles or dandruff etc. have fallen on the surface.	Do not leave the lid of the chamber open too long; apply the samples as quickly as possible; do not lean over the gel.
Smudged bands, the bands tail.	There is grease in the sample.	Remove lipophilic substances completely during sample preparation.
	Insufficient loading with SDS.	The sample buffer should contain at least 1% SDS; the SDS/sample ratio should be higher than 1.4:1.
	Incomplete stacking of some highly concentrated protein fractions.	Use a complete discontinuous buffer system either by casting a resolving and a stacking gel in the traditional way, or by equilibrating the stacking zone of a film supported gel selectively in a vertical buffer chamber (see page 187).
	Strongly acidic and basic proteins as well as nucleoproteins do not react well in the SDS system.	Try CTAB electrophoresis (cationic detergent in an acid buffer system) (Eley *et al.* 1979, Atin *et al.* 1985) .

The molecular weights do not agree with other experimental results.	The molecular weights of non-reduced samples were determined with reduced marker proteins. Glycoproteins migrate more slowly than polypeptides with the same molecular weight in the SDS buffer system used, because the sugar moieties are not charged with SDS.	An estimation of the molecular weights is possible by comparison with non-reduced globular polypeptides. Use a Tris-borate-EDTA buffer; borate binds to the sugar moieties and the mobility increases because of the additional negative charge. Also use a gradient gel.
Background in the separation lanes.	Protease activity in the sample.	Proteases are also active in the presence of SDS; add inhibitor if necessary (e.g. PMSF).

Tab. A2-4: SDS-PAGE specific staining problems.

Symptom	Cause	Remedy
Coomassie Blue: insufficient staining power.	SDS was not properly removed from the sample.	Only use pure SDS, low percentages of 14-C and 16-C sulfate bind more strongly to proteins; wash out SDS with 20% TCA; stain for longer than native electrophoresis.
	The alcohol content of the destaining solution is too high.	Reduce the ethanol or methanol content of the destaining solution, especially at high ambient temperatures; avoid alcohol containing staining solutions; or else: use colloidal development methods.
Silver staining: negative bands.	Impure SDS.	See above.

Tab. A2-5: Drying.

Symptom	Cause	Remedy
The gel comes off the support film during staining, tears during drying, the dried gel rolls up.	The bonds between the gel and the support film have been partly hydrolyzed by the strong acids (TCA, H_3PO_4) in some staining solutions, they do not resist the high mechanical tensions of the concentrated gels.	Prepare highly concentrated gels with a degree of crosslinking $C=2\%$ instead of $C=3\%$ as usual. But gels with less concentrated plateaus or stacking gels should be crosslinked with $C=3\%$ otherwise they are not stable enough; add 10% (v/v) glycerol to the last destaining solution, 15 min.

A3 Semi-dry blotting

Tab. A3-1: Assembling the blotting sandwich.

Symptom	Cause	Remedy
Air bubbles between the filter papers.	Air bubbles are trapped between the layers because the sandwich was not made up under buffer.	After assembling the sandwich carefully and slowly push out the air bubbles with a roller, trying not to expel too much buffer.
Gels which are polymerized on support films are difficult to handle.	It is difficult to remove the support film: the gels tear or warps.	Use the Film Remover.
Bits of the gel remain on the support film.	The metal wire was not pulled through regularly, the process was interrupted a few times.	Pull the wire completelythrough using a single action with constant speed .
The gel warps when the support film is removed.	Very thin gels or gels with large pores, as well as gels containing glycerol, often stick to the film.	Spray a a few drops of transfer buffer between the gel and the film with a Pasteur pipette.

Tab. A3-2: Problems during blotting.

Symptom	Cause	Remedy
The power supply does not work properly, it jumps between different current settings or switches itself off.	Some power supplies do not work at the very low voltages required for blotting: 3 to 6 V.	Use a power supply with a low resistance.
High power (ca. 20 W).	The surface of the filter paper is too large, the current flows around the blot stack.	Cut the filter paper to the size of the gel and the blotting membrane, or else cut a plastic mask and place it under the blotting membrane. (The window should be the same size as the gel and the blotting membrane).

The voltage increases during blotting.	Electrolysis gas pockets have formed between the graphite plates and the filter paper and this causes the conductivity to decrease.	Place a 1 kg weight on the graphite plate so that the trapped gas escapes from the sides.
The power increases during blotting, the blot sandwich warms up.	The buffer is too concentrated, the current is too high.	Only use the recommended buffer concentration, if possible do not set the current at more than 0.8 mA/cm^2.

Tab. A3-3: After blotting.

Symptom	Cause	Remedy
It is difficult to remove the blotting membrane.	The gel surface is too sticky.	Use a blotting membrane with 0.45 µm pores, no smaller, for gels with large pores (focusing gels, agarose gels); soak the gels briefly (3 to 4 min) in the transfer buffer before assembling the blotting sandwich; for IEF gels place the side of the gel that was previously bound to the support film on the blotting membrane (not the other surface).

Tab. A3-4: Results.

Symptom	Cause	Remedy
No transfer, there is nothing on the blotting membrane.	The current supply was wrongly connected.	Check the connections: the current should flow towards the anode in basic buffers and towards the cathode in acid buffers.
	The blotting sandwich was not correctly assembled, the blotting membrane is on the wrong side.	Check the assembling order of the blotting sandwich; see above.
	The surface of the filter paper is too large, the current flows around the blot.	Cut the filter paper and the blotting membrane to the size of the gel or else use a plastic mask, see above.

	The urea concentration of the focusing gel is too high: SDS does not bind to the proteins so the transfer does not occur because of the missing electrical charge.	Soak focusing gels which contain urea in the cathode buffer for a few minutes so that the urea diffuses out, then assemble the blot.
	The focusing gel contains non-ionic detergents which prevent the isoelectric proteins from becoming charged.	Native electrophoresis should be carried out in the absence of non-ionic detergents; as yet no satisfying solution exist for IEF (subject for a thesis).
	The proteins lie within the blotting membrane and cannot be seen.	Use a general stain to check, e.g. Amido Black or Indian Ink, then scan the blot with a laser densitometer.
The transfer is incomplete (the molecules stay in the gel).	The gel concentration is too high.	Use a lower gel concentration.
	The molecular weight of a few proteins is too high.	Carry out a limited proteolysis of the high molecular weight proteins before the transfer in the gel.
	The speeds of migration of the proteins are too variable: the high molecular weight ones are still in the gel, the low molecular weight ones have already passed through the membrane.	Use pore gradient gels.
	The methanol in the transfer buffer causes the gel to shrink too much.	Remove methanol or reduce its concentration.
Irregular transfer.	The current is irregular because the graphite plates are too dry.	Always wet the graphite plates with distilled water before assembly.
Incomplete transfer.	The transfer time was too short.	The recommended transfer time (1 h) should be prolonged (e.g. by 1/2 h) for thick gels, e.g. 3 mm and/or gels with very narrow pores.
	The charge/mass ratio of a few of the proteins is not favorable.	Equilibrate the gel in the cathode buffer for 5 to 10 min before blotting.

	Part of the current has flowed beside the gel and the blotting membrane.	Always cut the filter paper and blotting membrane to exactly the same size as the gel; or else use a plastic mask with openings to isolate the area around the gel.
	When several "trans units" are transferred simultaneously, the transfer efficiency decreases.	In such a case, only blot one transfer unit at a time.
	A continuous buffer system was used; this is less effective than a discontinuous buffer.	Use a discontinuous buffer system.
	The blotting with a discontinuous buffer system was performed at too low a temperature, this causes the pK value of the terminating ion to change and it migrates faster than the protein.	If 6-aminohexanoic acid is used as terminating ion, the blotting should be carried out at room temperature (20 to 25 °C); if the blotting of temperature sensitive enzymes has to be carried out at a lower temperature, use another buffer system.
Poor transfer (the pattern is distorted).	Contact problem: there are air bubbles in the blotting sandwich.	Make sure that there are no air bubbles; carefully press them out with a hand roller.
	The polyacrylamide gel swells during blotting.	Add 20% methanol to the transfer buffer; soak the gel in the transfer buffer beforehand (5 min).
	Diffusion.	Reduce the time for sandwich assembly and equilibration; do not shift the gel and the blotting membrane once they have come into contact.
	A lot of buffer in the graphite plates after long and /or intensive use of the semidry blotter	Always wet the graphite plates intensively with distilled water before assembling the blot. From time to time the buffer ions must be eluted from the graphite plates. Soak a stack of filter paper in blotting buffer and make a 30 min run with reversed polarity (change anode and cathode plug in the power supply sockets). Repeat this procedure two times. Soak tissue paper with dist. water and lie them on the graphite plates for a few min. Repeat this.

The transfer was not effective (the molecules migrate out of the gel, but too few are found on the membrane).	The binding is not strong enough.	After the transfer let the blot lie overnight or for 3 h in a heating cabinet at 60 °C so that the bonds are reinforced; only then stain or block.
	The low molecular weight peptides have been washed out during detection.	Use another blotting membrane: e.g. nitrocellulose with a smaller pore size, cyanobromide activated NC, PVDF or nylon film. Or else place several sheets of blotting membrane behind one another so that blotted molecules are trapped. Use nylon membranes and fix with glutaraldehyde after the transfer (Karey and Sirbasku, 1989).
	The binding capacity has been reduced by detergents.	Do not use detergents or use agarose-intermediate-gels (Bjerrum *et al.* 1987).
	The pH was too high or too low.	Modify the pH accordingly.
	The transfer time was too long.	Shorten the transfer time.
There are additional bands on the blotting membrane.	A few proteins were blotted through when several trans units were blotted simultaneously and have migrated on to the next membrane.	When several trans units are blotted simultaneously insert a dialysis membrane between them.
The transfer was not effective - the proteins have left the gel but are not on the blotting membrane.	The transfer time is too long; the proteins have been blotted through.	Shorten the transfer time; guideline: 1 h for IEF gels. Less time should be used for native gels.

Tab. A3-5: Detection.

Symptom	Cause	Remedy
Fast Green cannot be washed out.	The development was too long.	Stain for a shorter time.

Too little protein was detected.	The detergent (e.g. Tween, NP-40) in the blocking buffer has eluted part of the proteins. The sensitivity is too low.	Reduce the detergent concentration or do without detergent. The sensitivity of immuno detection can be increased by alkaline treatment; perform immuno reactions and enzyme reactions at 37 °C.
	The sensitivity of the peroxidase reaction is too low.	Use another peroxidase substrate, e.g. the tetrazolium method; use the immuno gold method.
	The sensitivity of the immuno gold reaction is too low.	Enhance it by silver staining.
	The sensitivity is very poor because the antigen concentration is extremely low.	Use an amplifying enzyme system: avidin-biotin-alkaline phosphatase method; or use an enhanced chemiluminescent detection method; or else use autoradiography.
Dark background.	Ineffective blocking.	Block for a longer time and at a higher temperature (37 °C).
	Cross-reactions with the blocking reagent.	Use another blocking reagent e.g. fish gelatin or skim milk powder.
The immuno detection is not specific.	SDS was not properly washed out, antibodies also bind to SDS protein micelles.	Wash longer.
	Poor quality antibodies. The wrong secondary antibodies were used. For example: investigation of antigens of plant origin. The secondary antibody animal nourishes itself with plants.	Use other antibodies. Use other secondary antibodies.
Indian Ink staining: too low sensitivity, some bands in the middle are not stained.	The Indian Ink solution was too alkaline.	Add 1% acetic acid to the Indian Ink solution.
The background is not completely white. No bands.	There are particles in the Indian Ink solution. The wrong dye was used.	Filter the Indian Ink solution, do not touch it with bare hands. Use fountain pen ink "Fount India".

A4 Two-dimensional electrophoresis (IPG-DALT)

Tab. A4-1: First dimension: IEF in immobilized pH gradients.

Symptom	Cause	Remedy
The urea in the IPG strips crystallizes out.	The IEF temperature is too low.	Focus at 15 to 20 °C.
	The surface dries out.	Add 0.5% Triton X-100 to the rehydration solution.
Sparking.	High voltage and ions in the contact fluid.	Use kerosene or DC-200 silicone oil.
Formation of strong precipitates at the sample application point.	There are aggregates and complexes in the sample.	Check the sample preparation procedure: see above. The addition of at most 0.8% (w/v) carrier ampholyte to the sample increases the solubility.
	The protein and/or salt concentrations in the sample are too high.	Dilute the sample, it is preferable to apply a higher sample volume. The contact surface between the sample and the gel should be as small as possible. Apply the samples with pieces of tubing or an applicator strip.
	The field strength is too high at first.	Do not prefocus (the pH gradient already exists); when IEF is carried out in individual strips, regulate the field strength with the voltage: 1 h at max. E=40 V/cm, then turn up the setting.
The bands in the basic area are blurred.	Influence of CO_2.	Catch off CO_2 during focusing; seal the separation chamber, add soda lime or 1 mol/L NaOH to the buffer tanks.
Horizontal streaks.	A few proteins have not focused.	Lengthen the focusing time, see above.
	IEF temperature too low.	Run the first dimension at 20 °C.
	A few proteins have precipitated on the surface of the IEF gel but dissolved again after a while.	Minimize precipitate formation (see above).

Horizontal streaking at the basic end.	Lack of reducing agent.	Place extra paper strips soaked with 15 mmol/L DTT on the surface of the IPG strips alongside the cathodal electrode strip Görg *et al.* (1995).
Unfocused (streaking) bands in the very basic part (pH > 10).	Unstable matrix, electro-endosmosis, CO_2 influence.	Prepare special IPG gels for very basic samples, rehydrate and run them acc. to Görg *et al.* (1996).
There are precipitates on the surface.	The proteins are not properly solubilized.	Increase the urea concentration (up to 9 mol/L); add a non-ionic or a zwitterionic detergent to the sample and the gel; add carrier ampholyte and DTT to the solubilizing solution. If this does not help, tra nondetergent sulfo-betains (Vuillard *et al.* 1995).
	The proteins concentrate before entering the gel and suddenly aggregate.	Use a low field strength at first. Mix the sample with a granular gel (Sephadex IEF) before application.
	Nucleic acids in the sample have precipitated with basic carrier ampholytes.	Apply the sample at the anode.
	High molecular weight nucleic acids form strongly ionic precipitates which bind to the protein.	Treat the sample with RNAse or DNAse.
	The basic part of the gradient is overbuffered by 2-mercaptoethanol (pK 9.5) from the sample.	Apply the sample at the anode, use DTT.

Tab. A4-2: Second dimension: SDS-electrophoresis.

Symptom	Cause	Remedy
The IPG strip becomes thinner at the anodic end and swells at the cathodic end and then turns up, the gel burns through.	Electroendosmosis: the IPG gel becomes negatively charged by equilibration in SDS, this induces transport of water towards the cathode.	Equilibration for 2×15 min under constant agitation in a modified equilibration buffer (add 6 mol/L urea, 30% glycerol) compensates the electro-osmotic effect; remove the IPG strips after 75 min, displace the cathodic electrode strips.
Spots are missing, loss of protein in the second dimension.	Electroendosmosis (see above): electroosmotic flow carries part of the proteins (up to 2/3) towards the cathode.	Use a modified equilibration system as described above for horizontal as well as vertical systems.
Horizontal streaks on the gel.	The proteins are not completely focused.	Focuse for a longer time; this is no problem during IPG because the gradient cannot drift.
	The equilibration was not effective enough.	Use a modified equilibration system as described above; the times $(2 \times 15$ min) must absolutely be held.
	Artifacts due to the reducing agent occur.	Add iodoacetamide (4 times the amount of DTT) to the second equilibration step (to trap the excess of reducing agent).
Horizontal streaks.	There are air bubbles between the first- and second-dimensional gels.	Ensure that the contact between the first- and second-dimensional gels is free from air bubbles.
There are three streaks over the whole width of the gel.	Artifacts due to the reducing agent.	Equilibrate in two steps; during the second step trap the excess of reducing agent with iodoacetamide.
Vertical stripes.	Problems with protein solubility.	Prepare a fresh urea solution to prevent the formation of isocyanate; use urea (6 mol/L) to equilibrate; increase the SDS content to 2%.
	Artifacts due to the reducing agent.	See above.

There are vertical streaks and a distorted spot pattern.	Micelles have formed between the non-ionic detergent from the IEF gel and the anionic detergent.	Only use 0.5% non-ionic detergent in the gel instead of 2% as usually used for IEF, or else use narrower strips for the first dimension.
There are vertical streaks in the high molecular weight range.	The equilibration step was not sufficient for several of the proteins.	Lengthen the equilibration time, increase the SDS concentration in the equilibration buffer (up to 4%), increase the temperature (up to 80 °C).
	The protein concentrating effect ("stacking") is not sufficient.	Use a discontinuous gel system.
There are vertical streaks in the low molecular weight range.	The protein concentration effect ("stacking") is not sufficient.	Use the method of Schägger and von Jagow (1987) for the second dimension.
There is a dark background in different areas of the gel.	There is protease activity in the sample.	Check the sample preparation method; if necessary add protease inhibitor (e.g. 8 mmol/L PMSF).
There is a conspicuous row of spots with the same moleular weight.	A few proteins were carbamylated by isocyanate.	Check the sample preparation procedure; prepare a fresh urea solution, avoid high temperatures; only use very pure urea.
Spots are missing.	The first dimension was stored too long or not correctly.	Equilibrate immediately after the first dimension and carry out the second dimension; or store the IPG strips in liquid nitrogen or at < −80 °C.
	Some proteins are not soluble any more after the first-dimension run has been stained or fixed.	Perform intermediate staining or fixing only of those proteins which are easily soluble.

A5 DNA electrophoresis

Tab. A5-1: Preparing the samples.

Symptom	Cause	Remedy
Inadequate resolution.	Sample too concentrated.	Thumb rule: When a sample can be detected with Ethidiumbromide, dilute it 1 : 5 with sample buffer.
SSCP analysis: Only double bands in the gel.	Chilling of the sample not efficient.	Use an ice-water bath instead of crushed ice. Apply samples quickly and start electrophoresis as fast as possible.
Mutation can not be detected.	SSCP analysis is influenced by many factors.	For a strategy to achieve optimal results check method 12 on page 243 ff.
Heteroduplex analysis: Only homoduplex bands and single strands in the gel.	Sample was too diluted to form heteroduplexes.	Samples must have the concentration to be detectable with Ethidiumbromide during heating and cooling. They are diluted to silver staining concentrations afterwards.
Denaturing run: Blurred bands an inadequate resolution.	Sample has not been denatured before application.	Heat sample with formamide before application.

Tab. A5-2: Running the gels.

Symptom	Cause	Remedy
Inadequate separation, poor resolution, no polymorphism expressed.	Wrong buffer system applied: different buffer systems can result in different patterns. Important for SSCP!	Test for optimal buffer system, apply: Tris-acetate / tricine; Tris-phosphate / borate, or else.
	Wrong temperature applied. Different temperatures can result in different patterns. Important for SSCP!	Test for optimal temperatures, run gel at different temperatures: 5 °C, 15 °C, 25 °C.

	Electrode solutions have been mixed up.	Add bromophenol blue to the anodal buffer to avoid this mistake.
	Electrodes are too close together; this leads to a lack of ions.	Set the electrodes on the outer edges of the electrode wicks, see drawing.
	Gel surface too wet.	Dry the gel surface with the edge of a filter paper until you can hear a "squeaking".
Curved front and band distribution.	Uneven rehydration of the gel.	Lift the edges of the gel-film repeatedly , or rehydrate on a rocking platform, use sufficient buffer volume.
Smiling effect.	Too much kerosene applied.	Apply only a very thin layer of kerosene with tissue paper to the cooling plate. A few air bubbles do not matter!
Uneven front and band distribution.	Air bubbles under electrode wick, uneven contact of gel edge and electrode wick.	Slide bent tip foreceps aong the edges of the wicks laying in contact with the gels.
Slanted front and band distribution.	Uneven buffer concentration in the wicks.	After soaking electrode wicks in buffer, hold them horizontally with two foreceps during the transfer from the PaperPool to the gel.
Sample does not stay in sample well, spreads out over stacking gel.	Water condensation on the gel due to high humidity and / or high room temperature.	Apply the sample on the gel, when cooling plate has still ambient temperature. Connect cooling plate to thermostatic circulator after sample application.
	Glycerol or sucrose, or organic solvent in sample, osmotic distribution.	Prepare sample without glycerol or sucrose; those are only necessary for submarine agarose gels and vertical PAGE. Remove organic solvent with a SpeedVac.
	Water condensation on the gel due to high humidity and / or high room temperature.	Apply the sample on the gel, when cooling plate has still ambient temperature. Connect cooling plate to thermostatic circulator after sample application.

| Less than 6 μL fit into sample well of a gel. | Buffer in sample wells: due to insufficient drying after rehydration. | Dry also the sample wells with filter paper. |
| | Buffer in sample wells: due to flooding with buffer from the cathodal electrode wick. | After soaking the cathodal wick: flap it upright along the long edge with two forceps, leave it for a few seconds to drain excess buffer off. Lay this wick on the cathodal edge of the gel with the formerly upper edge oriented towards the sample wells. |

Tab. A5-3: Silver staining.

Symptom	Cause	Remedy
No bands detected.	Formaldehyde has precipitated.	Store formaldehyde at room temperature, never in a refrigerator.
Gel shows dark background, and no or faint bands.	Compositon and/or quality of chemicals and/or water is inadequate.	Use only high purity chemicals. Try another supplier for $AgNO_3$.
	Plastic trays are used, which contain softener compounds, which reduce silver.	Use only glass or stainless steel trays. Clean these trays thoroughly after each silver staining.
Dark brown background.	Temperature of solutions to high.	Perform silver staining with solutions cooled down below + 15 °C.
Gel shows darker areas and white areas after silver staining.	Uneven rehydration of the gel.	Lift the edges of the gel-film repeatedly , or rehydrate on a rocking platform , use sufficient buffer volume.
Silver stained fragment can not be reamplified.	Wrong staining protocol. TCA and/or glutardialdehyde has destroyed the complete DNA in a band.	Use only DNA staining protocols, not those for proteins.

A6 Vertical PAGE

Only problems specific to vertical gels are listed here. Consult paragraphs A2 and A4 for general problems occuring in SDS and Two-dimensional electrophoresis.

Tab. A6-1: Gel preparation.

Cause	Cause	Remedy
Refractive lines in different directions are visible in the gel.	Dirty plates.	Wash and dry glass and ceramics plates thoroughly.
	CelloSeal®on the ceramics plate.	Do not coat the gaskets with CelloSeal®.
Sample well arms are too short.	Poor polymerization effectiveness.	Use stacking gel buffer pH 6.8 for the low acrylamide part; degas monomer solution, add more APS solution.
Stacking gel and well arms have dried.	Gels have not been kept in sufficient humidity.	Remove gels after 1 hour after casting from the stand and place them immediately with enough gel buffer into completely sealed plastic bags.

Tab. A6-2: Running the gels.

Symptom	Cause	Remedy
Cathodal buffer leaking out.	A crinkle in the gasket has formed.	Take the gasket out and place it back into the groove. Do not coat it with CelloSeal®.
Sample (visible because of added bromophenol blue) does not deposit at the bottom of the well.	No density difference. The samples must contain sucrose or glycerol.	Add 25 % (v/v) glycerol to the sample buffer.
Bromophenol blue dye does not migrate.	Incomplete filling of the cathodal compartment.	Fill 75 mL in, check buffer level, check gasket for leaking.
	Wrong buffer in the cathodal compartment.	Pour the Tris-glycine buffer into the cathodal compartment.

	One gel is run. Short circuit in the anode buffer.	Clamp a glass or a ceramics plate to the opposite side of the core.
"Smiling front".	Insufficient dissipation of Joule heat.	Reduce current setting. When direct cooling is employed: check, whether the Parafilm® has been removed before clamping the cassette to the core.

Tab. A6-3: Separation results.

Cause	Cause	Remedy
The bands are close together at the front.	Poor sieving properties of the gel.Gel has partially hydrolized due to: too long storage. too warm storage.	Store gels in a refrigerator, however no longer than 2 months.

Tab. A6-4: Preservation.

Cause	Cause	Remedy
Dried with cellophane clamped in two frames: Gels are cracking.	Air bubbles caught between gel and cellophane.	Carefully remove all air bubbles.
	Cellophane dries too fast, because it has been soaked in water.	Soak cellophane in the same glycerol solution like the gel.

References

Aebersold RH, Teplow D, Hood LE, Kent SBH. J Biol Chem. 261 (1986) 4229-4238.

Aebersold RH, Pipes G, Hood LH, Kent SBH. Electrophoresis. 9 (1988) 520-530.

Allen RC, Budowle B, Lack PM, Graves G. In Dunn M, Ed. Electrophoresis´86. VCH, Weinheim (1986) 462-473.

Altland K, Banzhoff A, Hackler R, Rossmann U. Electrophoresis. 5 (1984) 379-381.

Altland K, Hackler R. In: Electrophoresis´84. Neuhoff V, Ed.Verlag Chemie, Weinheim (1984) 362-378.

Altland K. Electrophoresis. 11 (1990) 140-147.

Alwine JC, Kemp DJ, Stark JR. Proc Natl Acad Sci USA. 74 (1977) 5350-5354.

Anderson NG, Anderson NL. Anal Biochem. 85 (1978) 331-340

Andrews AT. Electrophoresis, theory techniques and biochemical and clinical applications. Clarendon Press, Oxford (1986).

Ansorge W, De Maeyer L. J Chromatogr. 202 (1980) 45-53.

Ansorge W, Sproat BS, Stegemann J, Schwager C. J Biochem Biophys Methods. 13 (1986) 315-323.

Atin DT, Shapira R, Kinkade JM. Anal Biochem. 145 (1985) 170-176.

Baldo BA, Tovey ER. Ed. Protein blotting. Methodology, research and diagnostic applications. Karger, Basel (1989).

Baldo BA. In Chrambach A, Dunn M, Radola BJ. Eds. Advances in Electrophoresis 7. VCH, Weinheim (1994) 409-478.

Barros F, Carracedo A, Victoria ML, Rodriguez-Calvo MS. Electrophoresis 12 (1991) 1041-1045.

Bassam BJ, Caetano-Annollés G, Gresshoff PM. Anal Biochem. 196 (1991) 80-83.

Bauer D, Müller H, Reich J, Riedel H, Ahrenkiel V, Warthoe P, Strauss M. Nucleic Acid Res. 21 (1993) 4272-4280.

Baumstark M, Berg A, Halle M, Keul J. Electrophoresis. 9 (1988) 576-579.

Bayer EA, Ben-Hur H, Wilchek M. Anal Biochem. 161 (1987) 123-131.

Beisiegel U. Electrophoresis. 7 (1986) 1-18.

Berson G. Anal Biochem. 134 (1983) 230-234.

Bjellqvist B, Ek K, Righetti PG, Gianazza E, Görg A, Westermeier R, Postel W. J Biochem Biophys Methods. 6 (1982) 317-339.

Bjellqvist B, Linderholm M, Östergren K, Strahler JR. Electrophoresis. 9 (1988) 453-462.

Bjellqvist B, Sanchez J-C, Pasquali C, Ravier F, Paquet N, Frutiger S, Hughes GJ, Hochstrasser D. Electrophoresis. 14 (1993) 1375-1378.

Bjerrum OJ, Selmer JC, Lihme A. Electrophoresis. 8 (1987) 388-397.

Bjerrum OJ. Ed. Paper symposium protein blotting. Electrophoresis. 8 (1987) 377-464.

Blake MS, Johnston KH, Russell-Jones GJ. Anal Biochem. 136 (1984) 175-179.

Blakesley RW, Boezi JA. Anal Biochem. 82 (1977) 580-582.

Blomberg A, Blomberg L, Norbeck J, Fey SJ, Larsen PM, Roepstorff P, Degand H, Boutry M, Posch A, Görg A. Electrophoresis. 16 (1995) 1935-1945.

Blum H, Beier H, Gross HJ. Electrophoresis. 8 (1987) 93-99.

Bøg-Hansen TC, Hau J. J Chrom Library. 18 B (1981) 219-252.

Bosch TCG, Lohmann J. In Fingerprinting Methods based on PCR. Bova R, Micheli MR, Eds. Springer Verlag Heidelberg, in press.

Brada D, Roth J. Anal Biochem. 142 (1984) 79-83.

Brewer JM. Science. 156 (1967) 256-257.

Brown RK, Caspers ML, Lull JM, Vinogradov SN, Felgenhauer K, Nekic M. J Chromatogr. 131 (1977) 223-232.

Budowle B, Giusti AM, Waye JS, Baechtel FS, Fourney RM, Adams DE, Presley LA et al. Am J Hum Genet. 48 (1991) 841-855.

Burnette WN. Anal Biochem. 112 (1981) 195-203.

Bussard A. Biochim. Biophys Acta. 34 (1959) 258-260.

Caetano-Annollés G, Bassam BJ, Gresshoff PM. Bio/Technology 9 (1991) 553-557.

Chiari M, Casale E, Santaniello E, Righetti PG. Theor Applied Electr. 1 (1989a) 99-102.

Chiari M, Casale E, Santaniello E, Righetti PG. Theor Applied Electr. 1 (1989b) 103-107.

Chrambach A. The practice of quantitative gel electrophoresis. VCH Weinheim (1985).

Cohen AS, Karger BL. J Chromatogr. 397 (1987) 409-417.

Davis BJ. Ann NY Acad Sci. 121 (1964) 404-427.

Denhardt D. Biochem Biophys Res Commun. 20 (1966) 641-646.

Desvaux FX, David B, Peltre G. Electrophoresis 11 (1990) 37-41.

Diezel W, Kopperschläger G, Hofmann E. Anal Biochem. 48 (1972) 617-620.

Dockhorn-Dworniczak B, Aulekla-Acholz C, Dworniczak B. Pharmacia LKB Sonderdruck A37 (1990).

Dunn MJ, Burghes AHM. Electrophoresis. 4 (1983a) 97-116.

Dunn MJ, Burghes AHM. Electrophoresis. 4 (1983b) 173-189.

Eckerskorn C, Mewes W, Goretzki H, Lottspeich F. Eur J Biochem. 176 (1988) 509-519.

Eckerskorn C, Lottspeich F. Chromatographia. 28 (1989) 92-94.

Eckerskorn C, Strupat K, Karas M, Hillenkamp F, Lottspeich F. Electrophoresis 13 (1992) 664-665.

Eley MH, Burns PC, Kannapell CC, Campbell PS. Anal Biochem. 92 (1979) 411-419.

Everaerts FM, Becker JM, Verheggen TPEM. Isotachophoresis, Theory, instrumentation and applications. J Chromatogr Library Vol 6. Elsevier, Amsterdam (1976).

Ferguson KA. Metabolism. 13 (1964) 985-995.

Fischer SG, Lerman LS. Proc Natl Acad Sci. 60 (1983) 1579-1583.

Fujimura RK, Valdivia RP, Allison MA. DNA Prot Eng Technol. 1 (1988) 45-60.

Gershoni JM, Palade GE. Anal Biochem. 112 (1983) 1-15.

Giaffreda E, Tonani C, Righetti PG. J Chromatogr. 630 (1993) 313-327.

Gianazza E, Chillemi F, Duranti M, Righetti PG. J Biochem Biophys Methods. 8 (1983) 339-351.

Görg A, Postel W, Westermeier R. Anal Biochem. 89 (1978) 60-70.

Görg A, Postel W, Westermeier R, Gianazza E, Righetti PG. J Biochem Biophys Methods. 3 (1980) 273-284.

Görg A, Postel W, Günther S, Weser J. Electrophoresis. 6 (1985) 599-604.

Görg A, Postel W, Weser J, Günther S, Strahler JR, Hanash SM, Somerlot L. Electrophoresis. 8 (1987a) 45-51.

Görg A, Postel W, Weser J, Günther S, Strahler JR, Hanash SM, Somerlot L. Electrophoresis. 8 (1987b) 122-124.

Görg A, Postel W, Günther S. Electrophoresis. 9 (1988a) 531-546.

Görg A, Postel W, Weser J, Günther S, Strahler JR, Hanash SM, Somerlot L, Kuick R. Electrophoresis. 9 (1988b) 37-46.

Görg A. Nature .349 (1991) 545-546.

Görg A. Biochem Soc Trans. 21 (1993) 130-132.

Görg A. In: Celis J, Ed. Cell Biology: A Laboratory Handbook. Academic Press Inc., San Diego, CA. (1994) 231-242.

Görg A, Boguth G, Obermaier C, Posch A, Weiss W. Electrophoresis. 16 (1995) 1079-1086.

Görg A, Obermaier C, Boguth G, Csordas A, Diaz J-J, Madjar J-J. Electrophoresis. (1997) in press.

Grabar P, Williams CA. Biochim Biophys Acta. 10 (1953) 193.

Günther S, Postel W, Weser J, Görg A. In: Dunn MJ, Ed. Electrophoresis '86. VCH Weinheim (1986) 485-488.

Hanash SM, Strahler JR, Somerlot L, Postel W, Görg A. Electrophoresis. 8 (1987) 229-234.

Hanash SM, Strahler JR. Nature. 337 (1989) 485-486.

Hanash SM, Strahler JR, Neel JV, Hailat N, Melham R, Keim D, Zhu XX, Wagner D, Gage DA, Watson JT. Proc Natl Acad Sci USA. 88 (1991) 5709-5713.

Hancock K, Tsang VCW. Anal Biochem. 133 (1983) 157-162.

Handmann E, Jarvis HM. J Immunol Methods. 83 (1985) 113-123.

Hannig K. Electrophoresis. 3 (1982) 235-243.

Hayashi K, Yandell DW. Hum Mutat. 2 (1993) 338 - 346.

Hedrick JL, Smith AJ. Arch Biochem Biophys. 126 (1968) 155-163.

Heukeshoven J, Dernick R. In: Radola BJ, Ed. Electrophorese-Forum '86. (1986) 22-27.

Hjalmarsson S-G, Baldesten A. In: CRC Critical Rev in Anal Chem. (1981) 261-352.

Hjertén S.Arch Biochem Biophys Suppl 1 (1962) 147.

Hjertén S. J Chromatogr. 270 (1983) 1-6.

Hoefer Protein Electrophoresis Applications Guide (1994) 18-54.

Hoffman WL, Jump AA, Kelly PJ, Elanogovan N. Electrophoresis. 10 (1989) 741-747.

Hsam SLK, Schickle HP, Westermeier R, Zeller FJ. Brauwissenschaft 3 (1993) 86-94.

Hsu D-M, Raine L, Fanger H. J Histochem Cytochem. 29 (1981) 577-580.

Jackson P, Thompson RJ. Electrophoresis. 5 (1984) 35-42.

Jaksch M, Gerbitz K-D ,Kilger C. J Clin Biochem. 1995 in press.

Jeppson JO, Franzen B, Nilsson VO. Sci Tools. 25 (1978) 69-73.

Johansson K-E. Electrophoresis. 8 (1987) 379-383.

Johnson DA, Gautsch JW. Sportsman JR. Gene Anal Technol. 1 (1984) 3-8.

Jorgenson JW, Lukacs KD. AnalChem 53 (1981) 1298-1302.

Jovin TM, Dante ML, Chrambach A. Multiphasic buffer systems output. Natl Techn Inf Serv. Sprinfield VA USA PB(1970)196 085-196 091.

Karey KP, Sirbasku DA. Anal Biochem. 178 (1989) 255-259.

Keen JD, Lester D, Inglehearn C, Curtis A, Bhattacharya. Trends Genet. 7 (1991) 5.

Kittler JM, Meisler NT, Viceps-Madore D. Anal Biochem. 137 (1984) 210-216.

Klose J. Humangenetik. 26 (1975) 231-243.

Kohlrausch F. Ann Phys. 62 (1987) 209-220.

Kohn J. Nature 180 (1957) 986-988.

Krause I, Elbertzhagen H. In: Radola BJ, Ed. Elektrophorese-Forum'87. (1987) 382-384.

Kyhse-Andersen J. J Biochem Biophys Methods. 10 (1984) 203-209.

Laboratory Protocols for Mutation Detection. Landegren E, Ed. Oxford University Press (1996).

Lämmli UK. Nature. 227 (1970) 680-685.

Laing P. J Immunol Methods. 92 (1986) 161-165.

Lane LC. Anal Biochem. 86 (1978) 655-664.

Laurell CB. Anal Biochem. 15 (1966) 45-52.

Leifheit H-J, Gathof AG, Cleve H. Ärztl Lab. 33 (1987) 10-12.

Liang, P, Pardee AB. Science 257 (1992) 967-971.

Loessner MJ, Scherer S. Electrophoresis.13 (1992) 461-463.

Lohmann J, Schickle HP, Bosch TCG. Bio-Techniques 18 (1995) 200-202.

Maniatis T, Fritsch EF, Sambrook J. Molecular cloning. A laboratory manual. Cold Spring Laboratory (1982).

Matsudaira P. J Biol Chem. 262 (1987) 10035-10038.

Maurer RH. Disk- Electrophorese - Theorie und Praxis der diskontinuierlichen Polyacrylamid-Elektrophorese. W de Gruyter, Berlin (1968).

Maxam AM, Gilbert W. Proc Natl AcadSci USA. 74 (1977) 560-564.

Merill CM, Goldman D, Sedman SA, Ebert MH. Science. 211 (1981) 1437-1438.

Moeremans M, Daneels G, De Mey J. Anal Biochem. 145 (1985) 315-321.

Moeremans M, Daneels G, Van Dijck A, Langanger G, De Mey J. J Immunol Methods. 74 (1984) 353-360.

Moeremans M, De Raeymaeker M, Daneels G, De Mey. J. Anal Biochem. 153 (1986) 18-22.

Möller A, Wiegand P, Grüschow C, Seuchter SA, Baur MP, Brinkmann B. Int J Leg Med 106 (1994) 183-189.

Montelaro RC. Electrophoresis. 8 (1987) 432-438.

Mosher RA, Saville DA, Thormann W. The Dynamics of Electrophoresis. VCH Weinheim (1992).

Neuhoff V, Stamm R, Eibl H. Electrophoresis. 6 (1985) 427-448.

O'Farrell PH. J Biol Chem. 250 (1975) 4007-4021.

Olsson BG, Weström BR, Karlsson BW. Electrophoresis. 8 (1987) 377-464.

Olszewska E, Jones K. Trends Gen. 4 (1988) 92-94.

Orita M, Iwahana H, Kanazewa H, Hayashi K, Sekiya T. Proc Natl Acad Sci USA. 86(1989) 2766-2770.

Ornstein L. Ann NY Acad Sci. 121 (1964) 321-349.

Ouchterlony Ö. Allergy. 6 (1958) 6.

Pflug W, Laczko B. Electrophoresis. 8 (1987) 247-248.

Pharmacia LKB Development Technique File No 230 PhastSystem (1989).

Pharmacia LKB Offprint. Instructions for the preparation of gels for the PhastSystemTM (1988).

Poduslo JF. Anal Biochem. 114 (1981) 131-139.

Prieur B, Russo-Marie F. Anal Biochem. 172 (1988) 338-343.

Puers C, Hammond HA, Jin L, Caskey CT, Schumm JW. Am J Hum Genet 53 (1993) 953-958.

Rabilloud T, Valette, Lawrence JJ. Electrophoresis. 15 (1994) 1552-1558.

Radola BJ. Biochim Biophys Acta. 295 (1973) 412-428.

Raymond S. Weintraub L. Science. 130 (1959) 711-711.

Rehbein H. Electrophoresis. 16 (1995) 820-822.

Rehbein H, Mackie IM, Pryde S, Gonzales-Sotelo C, Perez-Martin R, Quintero J, Rey-Mendez M. Inf. Fischwirtsch. 42 (1995) 209-212.

Reiser J, Stark GR. Methods Enzymol. 96 (1983) 205-215.

Renart J, Reiser J, Stark GR. Proc Natl Acad Sci USA. 76 (1979) 3116-3120.

Rickwood D, Hames BD. Gel electrophoresis of nucleic acids. IRL Press Ltd (1982).

Riesner D, Steger G, Wiese U, Wulfert M, Heibey M, Henco K. Electrophoresis 10 (1989) 377-389.

Righetti PG, Drysdale JW. Ann NY Acad Sci. 209 (1973) 163-187.

Righetti PG. J. Chromatogr. 138 (1977) 213-215.

Righetti PG. In:Work TS, Ed. Burdon RH. Isoelectric focusing: theory, methodology and applications. Elsevier Biomedical Press, Amsterdam (1983).

Righetti PG, Gelfi C. J Biochem Biophys Methods. 9 (1984) 103-119.

Righetti PG, Wenisch E, Faupel M. J Chromatogr. 475 (1989) 293-309.

Righetti PG. In: Burdon RH, van Knippenberg PH. Ed. Immobilized pH gradients: theory and methodology. Elsevier, Amsterdam (1990).

Rimpilainen M, Righetti PG. Electrophoresis. 6 (1985) 419-422.

Robinson HK. Anal Biochem. 49 (1972) 353-366.

Rosengren A, Bjellqvist B, Gasparic V. In: Radola BJ, Graesslin D. Ed. Electrofocusing and isotachophoresis. W. de Gruyter, Berlin (1977) 165-171.

Rossmann U, Altland K. Electrophoresis. 8 (1987) 584-585.

Rothe GM, Purkhanbaba M. Electrophoresis. 3 (1982) 33-42.

Rothe G. Electrophoresis of Enzymes. Springer Verlag, Berlin (1994).

Salinovich O, Montelaro RC. Anal Biochem. 156 (1986) 341-347.

Sanger F, Coulson AR. J Mol Biol. 94 (1975) 441-448.

Sanguinetti CJ, Dias NE, Simpson AJG. BioTechniques. 17 (1994) 915-919

Schägger H, von Jagow G. Anal Biochem. 166 (1987) 368-379.

Schägger H, von Jagow G. Anal Biochem. 199 (1991) 223-231.

Scherz H. Electrophoresis. 11 (1990) 18-22.

Schickle HP. GIT LaborMedizin. 19 (1996a) 159-163.

Schickle HP. GIT LaborMedizin. 19 (1996b) 228-231.

Schumacher J, Meyer N, Riesner D, Weidemann HL. J Phytophathol. 115 (1986) 332-343.

Schwartz DC, Cantor CR. Cell. 37 (1984) 67-75.

Serwer P. Biochemistry. 19 (1980) 3001-3005.

Seymour C, Gronau-Czybulka S. Pharmacia Biotech Europe Information Bulletin (1992).

Shapiro AL, Viñuela E, Maizel JV. Biochem Biophys Res Commun. 28 (1967) 815-822.

Simpson RJ, Moritz RL, Begg GS, Rubira MR, Nice EC. Anal Biochem. 177 (1989) 221-236.

Sinha PK, Bianchi-Bosisio A, Meyer-Sabellek W, Righetti PG. Clin Chem. 32 (1986) 1264-1268.

Sinha P, Köttgen E, Westermeier R, Righetti PG. Electrophoresis. 13 (1992) 210-214

Smith MR, Devine CS, Cohn SM, Lieberman MW. Anal Biochem. 137 (1984) 120-124.

Smithies O. Biochem J. 61 (1955) 629-641.

Southern EM. J Mol Biol. 98 (1975) 503-517.

Strahler JR, Hanash SM, Somerlot L, Weser J, Postel W, Görg A. Electrophoresis. 8 (1987) 165-173.

Strupat K, Karas M, Hillenkamp F, Ekkerskorn C, Lottspeich F. Anal Chem. 66 (1994) 464-470.

Susann J. The Valley of the Dolls. Corgi Publ. London (1966).

Sutherland MW, Skerritt JH. Electrophoresis. 7 (1986) 401-406.

Suttorp M, von Neuhoff N, Tiemann M, Dreger P, Schaub J, Löffer H, Parwaresch R, Schmitz N. Electrophoresis 17 (1996) 672-677.

Svensson H. Acta Chem Scand. 15 (1961) 325-341.

Taketa K. Electrophoresis. 8 (1987) 409-414.

Terabe S, Otsuka K, Ichikawa K, Tsuchiya A, Ando T. AnalChem.64 (1984) 111-113.

Terabe S, Chen N, Otsuka K. In Chrambach A, Dunn M, Radola BJ. Eds. Advances in Electrophoresis 7. VCH, Weinheim (1994) 87-153.

Tiselius A. Trans Faraday Soc. 33 (1937) 524-531.

Tovey ER, Baldo BA. Electrophoresis. 8 (1987) 384-387.

Towbin H, Staehelin T, Gordon J. Proc Natl Acad Sci USA. 76 (1979) 4350-4354.

Vandekerckhove J, Bauw G, Puype M, Van Damme J, Van Montegu M. Eur J Biochem. 152 (1985) 9-19.

Vesterberg, O. Acta Chem. Scand. 23 (1969) 2653-2666.

Vidal-Puig A, Moller DE. BioTechniques 17 (1994) 492-496.

Vuillard L, Marret N, Rabilloud T. Electrophoresis. 16 (1995) 295-297.

Wagner H, Kuhn R, Hofstetter S. In: Wagner H, Blasius E. Ed. Praxis der elektrophoretischen Trennmethoden. Springer-Verlag, Heidelberg (1989) 223-261.

Weber K, Osborn M. J Biol Chem. 244 (1968) 4406-4412.

Welsh J, McClelland M. Nucleic Acids Res. 18 (1990) 7213-7218.

Wenger P, de Zuanni M, Javet P, Righetti PG. J Biochem Biophys Methods 14 (1987) 29-43.

Westermeier R, Postel W, Weser J, Görg A. J Biochem Biophys Methods. 8 (1983) 321-330.

Westermeier R, Schickle HP, Thesseling G, Walter WW. GIT Labor-Medizin. 4 (1988) 194-202.

Westermeier R: In Doonan S. Ed. Protein Purification Protocols. Methods in Molecular Biology 59. Humana Press, Totowa, NJ (1996) 239-248.

Williams JGK, Kubelik AR, Livak KJ, Rafalski JA, Tingey SV. Nucleic Acids Res. 18 (1990) 6531-6535.

Willoughby EW, Lambert A. Anal Biochem. 130 (1983) 353-358.

White MB, Carvalho M, Derse D, O´Brien SJ, Dean M. Genomics 12 (1992) 301-306.

Index

A, C, G, T, see adenine, cytosine, guanine, thymine
Absorbance   87
Absorbance spectrum   87
ACES, see N-2-acetamido-2-amino-ethanesulfonic acid
Acrylamide   11, 95, 277, 288, 294, 297
– formula   11
– gel recipe, DNA gels   230, 233
– – IEF gel   153, 277
– – Immobiline gel   202, 217
– – native PAGE   135
– – SDS gradient gel   172, 219
– – titration curve gel   122
– – ultrathin layer electrophoresis   102
– – vertical gels   264ff
– gradient   30
– immobilized pH gradient   52
– polymerization   131, 151, 163, 230
Additive gradient   207, 251ff
Adenine   20
Adsorption, on blotting membranes   59, 63, 82
– on capillaries   7
A/D transformer, see analog digital transformer
Affinity electrophoresis   15, 36, 107
AFLP   255, 259
Agarose gel   9ff, 17ff, 47, 285
– blotting   60
– clinical routine   15
– electrophoresis   12
– immunoelectrophoresis   13
– native immunoblotting   69
– PFG electrophoresis   19
– preparation   107ff
– protein detection   117ff, 151ff
– rehydratable   48
– submarine chamber   17, 75
– submarine electrophoresis   17, 75
– transfer buffer   64 ff, 188
Agarose IEF   47

– method   143ff, 285ff
Alkaline blotting   66
Alkaline phosphatase   15
Alkylation   168, 300
Allelic ladder   27
Amido Black staining   193
Ammonium persulfate   10, 95, 163, 212, 277, 294, 298
– gel recipe, DNA gels   230, 233
– – IEF gel   153
– – Immobile gel   202, 217
– – native electrophoresis   135
– – SDS gradient gel   175, 219
– – titration curve gel   122
– – ultrathin layer gel electrophoresis   101
– – vertical gels   264ff
– polymerization   131, 143, 151, 163
– stock solution   95, 102, 119, 132, 152, 170, 197, 215, 230, 264
Amphoteric buffer   4, 35, 125, 131
Amphoteric substance   3, 35, 45, 56
Amplification products   23 ff, 231 ff, 241ff
Amplifying enzyme detection system   68
Analog digital transformer   7, 85
Analysis, quantitative   2, 42 ff, 81ff, 184ff, 257
Anion   3, 28, 42
Antibody   13, 59, 90, 114, 116
– immunoblotting   67, 302
– detection on blots   67f
Antibody, quantity   82
Antibody, reactivity   66
Antibody, solution   149
Antigen   13, 59, 114
Antigen-antibody titer   82
APS, see ammonium persulfate
ARDRA   24
Automation   7, 20, 23, 68, 71, 79
Autoradio-fluorography   85
Autoradiography   8, 19, 22, 37, 66, 258
– densitometry of –   90

BAC, see bisacrylylcystamine
Base pairs   22
Base pair ladder   129
Baseline   88
Bi-directional electrophoresis   19
Bi-directional transfer   60, 62
Biotin-avidin   20, 68, 307
Bis, see NN'-methylenebisacrylamide
Bisacryloylcystamine   11
Blocking   66, 307
Blotting   2, 8, 59ff, 81
– blotting equipment   60 ff, 79, 187ff
– electrophoretic   61, 183, 187ff, 302
– of IEF   60
– of Immobiline gels   211
– of SDS electropherograms   180, 187
– semi dry   62, 187, 302
– trouble-shooting   302ff
Blotting membranes   59ff, 94, 187ff, 194
– blocking of –   66
– densitometry of –   86
– staining of –   66, 19
Blue toning   179
bp, see base pairs
Bovine serum albumin   66, 90
BSA, see bovine serum albumin
Buffer   4, 28 ff, 64
– amphoteric   4, 35, 131
– agarose   78
– strips, paper   77, 105, 139, 176, 223f,
   236f, 255f, 260ff
– strips, polyacrylamide   34
– system, discontinuous   2, 28 ff, 65, 188,
   231, 296

C, see cross-linking
Calibration curve   33, 48, 89, 161, 183
Capillary blotting   60
Capillary electrophoresis   1, 6, 43, 81
Capillary isotachophoresis   43
CAPS, see 3-(cyclohexylamino)-propane-
   sulfonic acid
Carboxymethyl   64
Carrier ampholytes   48, 55ff, 277, 285
– recipe   123, 145, 155, 206, 218
Catalyst   10, 35, 119, 131, 143

Cathodic drift   51, 147, 162, 279
– see also plateau phenomenon
Cation   3, 28, 35, 42
CDGE see
Cellulose acetate   8, 13, 84
Cerebrospinal fluid   164
Cetyltrimethylammonium bromide   35,
   300
CHAPS, see 3-(3-cholamidopropyl)dime-
   thylammonio-1-propane sulfate
Charge heterogeneity   16, 81
Chemicals, list   95
3-(3-Cholamidopropyl)dimethylammonio-
   1-propane sulfate   164, 214, 218
Chromosome separation   18
Clinical diagnostic   12, 15, 107
Clinical laboratory   12
Clinical routine   8, 84
CM, see carboxymethyl
CMW, see Collagen Molecular Weight
Collagen Molecular Weight   89, 97, 167
Colloidal gold   66, 68
Colloidal staining, Coomassie   126, 140,
   157, 179, 209, 291
Complex formation   3, 82
Computer, for densitometric evaluation
   39, 83ff, 160ff, 181ff
– for DNA sequencing   21
Concentration effect   3
– see also zone sharpening effect
Configurational changes   3, 45, 47
Constant denaturing gel electrophoresis
   26, 249ff, 257ff
Consumables   92, 94
Contact fluid   190, 279, 291, 296, 308
– electrophoresis   105, 139, 176, 223,
   237, 255, 260
– immunoelectrophoresis   112
– isoelectric focusing   146, 156, 208, 220
– titration curve analysis   124
Continuous free flow zone electrophoresis
   5
Cooling   23, 71ff, 273, 242ff
– thermoelectric   76f
Coomassie Blue,   8, 35
– problems   291, 301

Coomassie staining   8, 293
– (Agarose IEF)   148
– colloidal   126, 140, 157, 179, 209, 293
– fast   126, 140, 158, 178, 223
– (SDS electrophoresis)   178
Coomassie staining (agarose electro-
   phoresis)   115
Coomassie staining (immunoelectro-
   phoresis)   116
Counter-ion   41
Counter immunoelectrophoresis   14
Creatinkinase   12
Cross-linking [%], degree of   11, 244
Cryo-IEF method   48
Cryo-isoelectric focusing   48, 164
Cryoproteins   48
CTAB, see cetyltrimethylammonium
   bromide
3-(Cyclohexylamino)-propanesulfonic
   acid   64
Cy5 label   21, 242
Cytosine   20, 21

Da, see Dalton
Dalton   15, 33, 89
DBM, see diazobenzyloxymethyl
cDNA   24
DDRT, see Differential display reverse
   transcription
DEAE, see diethylaminoethyl
Denaturing conditions
– proteins   31, 36, 47, 176ff, 164, 214ff
– DNA   19, 22, 26ff, 247ff, 257ff
Denaturing gradient gel electrophoresis
   26, 249ff
Denaturation   3, 19, 25, 165, 214, 243,
   258,
Densitogram, Densitometer,
Densitometry, Densitometric evaluation
   84ff, 159ff, 181ff
Density gradient   30, 53, 172, 201, 207,
   252
Desoxyribonucleic acid   16, 19ff, 61, 309,
   312f
– blotting   61ff
– separation methods   17ff, 231ff

Detection   7, 8, 14, 43, 67, 81ff
– see also autoradiography
– see also blotting
– see also staining
Detection, limit of   8, 12, 22, 23
Detector   6, 43
Detergent   7, 31ff, 240, 258, 260
– anionic   26, 165
– cationic   35, 300
– non-ionic   35, 36, 47, 260, 284, 292,
   309
– – for blotting   63
– – sample preparation   35, 214, 284,
   292, 309
– – rehydration   47, 162, 206, 218, 284,
– zwitterionic   48, 214, 218, 284, 309
Dextran gel   55, 215
DGGE see Denaturing gradient gel electro-
   phoresis
Diagnostic, clinical   12, 15, 107
Dialysis   173, 278, 306
Diazobenzyloxymethyl   63
Diazophenylthioether   63
Diethylaminoethyl   64
Differential display reverse transcription
   24f, 229ff, 257ff
Different light refraction   5
Diffuse   14, 30, 114
Diffusion   3, 8, 50, 283, 305
– immunodiffusion   14, 114
Diffusion blotting   60
Diffusion coefficient   50
Dimethylsulfoxide   162, 282
Disc., see discontinuous
Discontinuous buffer system   2, 40ff, 81,
   185
– blotting   65, 188, 305
– DNA separations   229ff, 244, 250,258
– proteins   28, 138, 170ff, 214ff, 264
Discontinuous electrophoresis   17, 28,
   131, 165, 263
Dissociation constant   1, 48, 52, 186, 217,
   196ff
Disulfide bond   11, 31, 165ff
Dithioerythritol   206
Dithiothreitol   31, 167, 206, 212, 216, 218

DMSO, see dimethylsulfoxide
DNA sequencing   19ff, 74
– automated   20
DNA, see desoxyribonucleic acid
Double Replica Blotting   62
DPT, see diazophenylthioether
DSCP   26, 129
DTE, see dithioerythreitol
DTT, dithiothreitol

EDTA, see ethylendinitrilotetracetic acid
Electric field   1, 18, 49
– strength   20, 41
Electroendosmosis   4, 9, 10, 17, 131, 240
– during counter immunoelectrophoresis
    14
– strong   286 ,310
Electroendosmosis, free of   143
Electroendosmotic effect   4, 48, 212, 278,
    286, 291, 310
– in capillaries   7
Electro-osmotic movement   4, 14
Electrode buffer   4, 12, 28f, 101,105, 220
– recipes for DNA   230, 247, 250, 259
– recipes for proteins   107, 111, 138, 169,
    265
Electrode solutions   50, 146f, 163, 281,
    286
Electropherogram   83
Electrophoresis   2, 7, 72
– automated   6, 20, 78f
– bidirectional   22
– continuous free flow zone   5
– discontinuous   28
– 2D electrophoresis   36, 213, 308
– in pulsed electric field   18, 76
2D electrophoresis, see two-dimensional
    electrophoresis
Electrophoretic blotting   61, 77, 187ff, 302
Electrophoretic elution method   55
Electrophoretic mobility   1ff, 6, 8, 12, 16,
    18, 41
Electrophoretic separation   4 , 29
Elution, electrophoretic   55
Enzyme, enzyme blotting   68

– enzyme detection system, amplifying
    68
– enzyme inhibitor   15, 54
– enzyme detection   8, 13, 15, 67
– enzyme substrate complex   48, 163
– enzyme substrate coupling reaction   8
– enzyme substrate reaction   67
– separation   8, 13, 15, 33, 163
Equilibration
– of gels   65, 190, 305
– of IPG strips   213, 216, 222, 309f
Equipment   73, 91f
Equivalence point   13
Ethidium bromide   17, 22. 84, 229, 242,
    276
Ethylendinitrilotetraacetic acid   19, 60,
    91, 230, 240, 250, 258, 276
– sample preparation   166, 169, 216, 233,
    245, 264, 292
Evaluation,   81ff, 128f
– densitometric   84ff, 160ff, 181
Extinction   86ff
– see also absorption

Fast Coomassie staining   126, 140, 158,
    178, 223, 276
Field strength   1, 20, 28, 41, 46, 71, 244,
    247
– setting   195, 209, 222, 232, 244
Fluorography   8, 20, 66
Focusing, isoelectric, see isoelectric
    focusing

GC, see group specific components
GC clamps   250
Gel   9
– non-restrictive   69
– rehydrated   35
– – for DNA   229ff, 246, 248, 249ff,
    258ff
– – for proteins   47, 137ff, 151, 162,
    217ff
– restrictive   16, 28, 69
Gel plate, horizontal   9, 22ff
Glass fiber membrane   56, 64
Glycoproteins   35, 63, 65, 68, 301

Gold, colloidal 66, 68
Gradient 3
Gradient gel 26, 30f, 45, 77
– preparation 172ff, 201ff, 249ff, 246, 294
Group specific component 206
Guanine 20, 21

Henderson-Hasselbach equation 53
HEPES, see N-(2-hydroxyethyl)piperazine-N'-2-ethanesulfonic acid
Heteroduplex 26, 229ff, 245
High Molecular Weight 97, 168
High Performance Capillary Electrophoresis 6, 82, 228
High Performance Liquid Chromatography 6, 82, 228
High-resolution 2D electrophoresis 36ff, 213ff
– trouble shooting 308ff
Histones 35, 228
HMW, see High Molecular Weight
HPCE, see High Performance Capillary Electrophoresis
HPLC, see High Performance Liquid Chromatography
Hybridization 67, 82
Hydrolysis 34, 196
Hydrophobic – interactions 36
– matrix 11
– proteins 32, 35, 36, 47
N-(2-Hydroxyethyl)piperazine-N'-ethanesulfonic acid 35, 97, 137ff

Identification 7, 67, 68, 228
IEF marker 48, 57, 97, 162
IEF, see isoelectric focusing
IgG, see immunoglobulin G
Immobiline 52ff, 55, 97, 196ff, 206, 288ff
– recipes 197ff, 217
Immobilizing membranes 59ff, 82
Immobilized pH gradients 38, 47, 52ff, 68, 195f, 288fff
– during 2D electrophoresis 38, 214
Immunoblotting 67, 90, 193, 307
– native 69

Immunodiffusion 14, 114
Immunoelectrophoresis 13ff, 36, 82, 107ff
Immunofixation 12, 107, 115, 143, 148, 164
Immunoglobulin G 27, 67, 114, 164
Immunoglobulin M 48
Immunoprecipitation 8, 13, 114ff, 148, 164
Immunoprinting 8, 13
Indian Ink staining 307
Integration 88, 160, 183
Interactions, intermolecular 3, 36
Instrumentation 71ff, 91
Ion-exchange chromatography 130
Ionic strength 4, 28, 33, 53, 72, 185
IPG, see immobilized pH gradients
IPG-Dalt 38, 213
– see also 2D electrophoresis: IPG/SDS electrophoresis
Iso-Dalt, see 2D electrophoresis: IEF/SDS electrophoresis
IsoDalt-system 37
Isoelectric focusing 2, 45ff, 129, 143ff, 151ff, 195ff
– blotting of - 60, 65, 303
– current during - 72
– during 2D electrophoresis 36ff, 213ff
– equipment 73ff, 91
– in agarose gel 143ff
– in capillaries 7
– in immobilized pH gradients 195f, 221ff
– in polyacrylamide gel 151ff
– principles 45ff
– sample application 82, 148, 156, 208, 218, 221f
– trouble-shooting 277ff
Isoelectric point 3, 46, 56f, 82, 160f
"Isoelectric ruler" 162
Isotachophoresis 2, 7, 29, 41ff
Isotachophoresis effect 62
ITP, see isotachophoresis

Joule heat 4, 6, 72, 74

Kerosene, see contact fluid

Laboratory, clinical   12
Laboratory equipment   91ff
Lactate dehydrogenase   13, 16
Laser   21, 68, 87, 91, 228
Laser densitometry   87, 160ff, 181ff
Lauryldimethylamine-N-oxide   162
LDAO, see Lauryldimethylamine-N-oxide
Leading ion   2, 28, 41, 188
Lectin   15, 57, 59
Lectin blotting   68, 90
Light refraction, different   5
Linear polyacrylamide   7
Lipopolysaccharide   8, 65
LMW, see Low Molecular Weight
Low molecular Weight   49, 97, 168
– peptides   34, 81, 186, 273ff
– substances   15, 101

Marker, mixture   161
– protein   33, 48, 82, 89ff, 97, 169
– DNA   229
Marking
– enzyme -, alkaline phosphatase   67, 307
– – peroxidase   67
– fluorescence   8, 20, 66f, 242
– non-radioactive   20, 67
– radioactive   8, 19, 67f, 258
Marking, enzyme, see also secondary antibody
Measurement, photometric   84ff, 160, 181
Medium, non-restrictive   12
– restrictive   12
– stabilizing   4, 8
MEKC see Micellar electrokinetic chromatography
Membrane,
– cellulose acetate   8
– immobilizing   59ff, 82
– isoelectric   55f
– proteins   35, 206, 214
MES, see 2-(N-morpholino)ethanesulfonic acid
Methyl cellulose   7, 228
NN'-Methylenebisacrylamide   11, 97, 277, 288, 294
– formula   11

– gel recipe, DNA gels   230
– – IEF gel   153, 227
– – Immobiline gel   202, 217
– – native PAGE   135
– – SDS gradient gel   172, 219
– – titration curve gel   122
– – vertical gels   264
– – ultrathin layer gel   101
– polymerization   131, 151, 163
Micellar electrokinetic chromatography   7
Migration, electrophoretic   14, 71
– speed of   71, 188
Mobility   28, 41, 47
– electrophoretic   1ff, 6, 8, 12, 18
– relative   7, 16
Molecular radius   16
Molecular weight   7, 31, 36, 83, 78, 92, 168, 183
– automatic determination   89, 182
– low   49, 97, 34, 81, 186, 273ff
– range   185
– separation   8, 31, 101ff, 131ff
Molecule-detergent micelle   4, 7, 31, 167ff
MOPS, see 3-(N-morpholino)propanesulfonic acid
2-(N-Morpholino)ethanesulfonic acid   35
3-(N-Morpholino)propanesulfonic acid   35
Moving boundary electrophoresis   5
m_r, see relative electrophoretic mobility
MW, see molecular weight

NAP, see Nucleic Acid Purifier
Native
– conditions   12, 22f, 35, 47, 67, 165
– electrophoresis   12, 16, 35, 107, 119, 131ff
– – in amphoteric buffers   4, 35, 56f, 131
– immunoblotting   67, 69
Net charge   3, 12, 30, 33
– isoelectric focusing   45f
– speed of migration   16ff, 71
Net charge curve   45, 57, 129f
Neutral blotting   66
Nitrocellulose   63, 194
Non-ionic detergent   31, 35, 47, 69, 98,

– in the gel   151, 278, 309
Non-restrictive medium   12
Nonidet, see non-ionic detergent
Nucleic Acid Purifier   92
– see also sample preparation
Nucleoprotein   35, 46
Nylon membrane   64, 66

O.D., see optical density
Oligoclonal IgG   164
Optical density   87
– see also extinction
Oxidation   162, 212, 284, 299

PAG, see polyacrylamide gel
PAGE, see polyacrylamide gel electrophoresis
PAGIEF, see polyacrylamide gel isoelectric focusing
Paper electrophoresis   8
Paper replica   55
PBS, see Phosphate Buffered Saline
PCR®, see Polymerase Chain Reaction
PEG, see polyethylene glycol
Peptide   12, 34, 55, 81, 186, 273f
– detection   195, 276
PFG, see Pulsed-Field Gel Electrophoresis
PGM, see phosphoglucomutase
pH gradient   2f, 46ff, 49, 56f, 72, 277ff
– 2D electrophoresis   36, 213ff, 308f
– IEF method   143, 151, 195, 197
– immobilized   38, 52ff, 68, 195ff, 288ff
Phenylmethyl-sulfonyl fluoride   98, 196, 214, 284, 301
Phosphate Buffered Saline   193
Phosphatase, alkaline   15
Phosphoglucomutase   206
Photometric measurement   84
pI, see isoelectric point
PI, see Protease Inhibitor
pK value, see dissociation constant
Plastic embedding of blotting membranes   194
Plateau gel   172, 219
– see also stacking gel
Plateau phenomenon   51, 279

PMSF, see phenylmethyl-sulfonyl fluoride
PMW, see Peptide Molecular Weight
Point, isoelectric   3, 45, 83, 89, 160,
Polyacrylamide gel   10ff, 15, 19ff, 184
– electrophoresis   35, 101, 263ff
– – DNA   19ff, 23ff, 229ff
– – proteins   120ff, 131ff,
– isoelectric focusing   47, 151ff, 195ff, 277ff
– linear   7
– rehydratable   151ff, 195ff, 213ff, 263ff, 277ff
Polyethylene glycol   162, 196
Polymerase Chain Reaction   22,ff, 229ff, 241ff, 249ff, 257ff
Polymerization   10, 52
– gels   104, 122, 135, 153f, 175, 204, 264ff
– catalyst   10, 35, 131, 175
– membranes   56, 194
Polyphenols   12, 101
Polysaccharide   8, 9
Polyvinyledendifluoride   63, 257
Pore gradient gels   30f, 88, 172ff. 267f, 271f, 304
Precipitation line   13, 14, 82
Pressure blotting   60
Protease Inhibitor   54, 206, 214
Protein A, radioiodinated   67
Protein sequencing   68, 228
Pulsed-Field Gel Electrophoresis   18
Purity control   81
Purity test   3
PVC, see polyvinyl chloride
PVDF, see polyvinylidenfluoride

Quantification   3, 81ff
Quantitative analysis   2, 42

r, see molecular radius
Radioiodinated protein A   67
RAPD, see Random amplified polymorphic DNA
Random amplified polymorphic DNA   23, 229ff
Rate of diffusion   46

RC, see retardation coefficient
Recipe, Immobiline   197
Rehydratable agarose gel   48
Rehydrated polyacrylamide gels
– DNA   229ff, 243ff, 249ff, 257ff
– proteins   47, 119ff, 131ff, 151ff, 195ff, 219
Relative distance of migration   16, 89
Relative electrophoretic mobility   1
Relative mobility   7, 16
Resolution   8, 46, 53, 185, 195, 273
Resolving gel   28ff, 263ff
Restriction Fragment Long Polymorphism   17
Restrictive medium   12, 16ff, 29
Retardation coefficient   16
Reversible staining   66, 193
Rf value, see m_r, R_m
– see relative electrophoretic mobility
Ribonucleic acid   16, 22, 53, 54
RFLP, see Restriction Fragment Long Polymorphism
R_m, see relative electrophoretic mobility
RNA, see ribonucleic acid
RNase protection assay   267ff
Rocket technique   14, 76
Round gel   31
Round gel tube   9
Routine, clinical   8, 77
RPA, see RNase protection assay

Sample   3
Sample application   156, 177, 208, 283
Sample preparation, proteins   76, 95, 284, 292, 310
– – denaturing   165, 214, 308
– – immunoelectrophoresis   107
– – isoelectric focusing   143, 151, 196, 278
– – native electrophoresis   132
– – SDS treatment   165ff, 298ff
– – titration curve analysis   119
– DNA   234, 245, 250, 258
– vertical gels   264
Schlieren optics   5
SDS, see sodium dodecyl sulfate

SDS, polyacrylamide electrophoresis   31ff, 36ff, 61, 81, 288ff
– gel   165ff, 213, 263ff
– pore gradient gel   187, 214, 267ff
SDS treatment   165, 264
Secondary antibody   68, 307
Semi dry blotting   62, 65, 77, 79, 187, 302
Separator IEF   51
Silicone oil, see contact fluid
Silver amplifying   68
Silver staining   22ff, 23, 71, 167, 263, 235, 253
– dried gels   116, 127, 141, 149, 158, 210
– recipes
– – DNA gels   240f, 314
– – protein gels   149, 159, 180, 223ff
Single strand conformation polymorphism   25, 243ff, 312
Slot former   102, 108, 133, 170, 231
Sodium dodecyl sulfate   31ff, 64, 69
Speed of migration   71, 188
SSCP, see Single strand conformation polymorphism
Stabilizing medium   4, 8
Stacking gel   28f, 34, 131ff, 165ff, 263ff
Staining of blotting membranes   66ff, 193
– reversible   66, 193
Staining, in stabilizing medium   8
– of blotting membranes   66
– see also detection
Starch gel   9
"Submarine" chamber, gel   17, 75
Surface charge   6

T, see total acrylamide concentration [%]
Tank blotting   62, 64
TBE, see Tris borate EDTA
TCA, see trichloro acetic acid
TEMED, see N,N,N';N'-tetramethylethylendiamine
Temperature gradient gel electrophoresis   26, 257ff
Terminating ion   2, 28, 42, 188
N,N,N'N'-Tetramethylethylendiamine   10, 35, 99, 143, 204, 288
– gel recipes   151, 153

– – DNA gels   130ff
– – Immobiline gel   201, 217
– – native electrophoresis   135
– – SDS gradient gel   172, 219
– – titration curve gel   122
– – ultrathin layer gel electrophoresis   98
– polymerization   131
– vertical gels   264ff
TGGE, see Temperature gradient gel electrophoresis
TF, see transferrin
Thermoblotting   59
Thin layer electrophoresis   8
Thymine   20, 21
Titration curve analysis   56ff, 119, 131, 284
TMPTMA, see trimethylolpropane-trimethacrylat [2-ethyl-2(hydroxymethyl)-1,3-propandiol-trimethacrylate]
Total acrylamide concentration [%]   11, 16
Transferrin   206
Trichloro acetic acid   99
– fixing   115, 126, 140, 148, 157,   210
Tricine, see N-tris(hydroxymethyl)-methyl-glycine
Trimethylolpropane-trimethacrylat [2-ethyl-2(hydroxymethyl)-1,3-propandiol-trimethacrylate]   99, 194
N-Tris(hydroxymethyl) aminomethane   65, 99, 188, 206, 212
– acetate   34, 185, 276

– ascorbic acid   212
– borate-EDTA   19, 35, 230, 246, 250, 258, 276, 301
– glycine   28ff, 64ff, 170,   188, 248
– HCl   28, 120, 132, 152, 166,   169, 219
– phosphate   28, 230, 246, 258
– Tricine   34f, 107, 186
– Tricine lactate   107
Tris, see tris(hydroxymethyl)-aminomethane
Triton X-100, see detergent, non-ionic
Trouble-shooting guide   277
Two-dimensional electrophoresis   36, 83, 213ff, 275, 312

Ultraviolet light   7, 43, 66, 81, 93, 194
Urea   19, 26ff, 36f, 47, 65, 69, 143, 249ff, 257ff
– 2D electrophoresis   36, 213
UV, see ultraviolet light

Vacuum blotting   61
Very Low Density Lipoproteins   206
Viscoelastic relaxation time   18
VLDL, see Very Low Density Lipoproteins

Zone electrophoresis   2, 12, 14, 46, 128
Zone sharpening effect   41

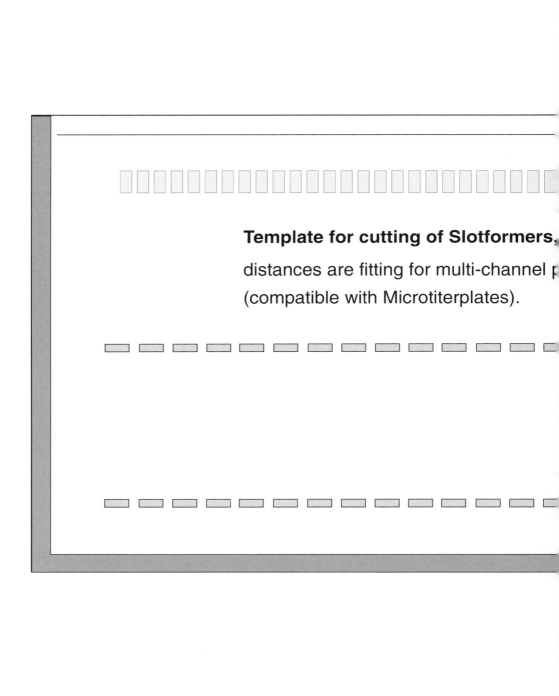

Template for cutting of Slotformers,

distances are fitting for multi-channel p

(compatible with Microtiterplates).

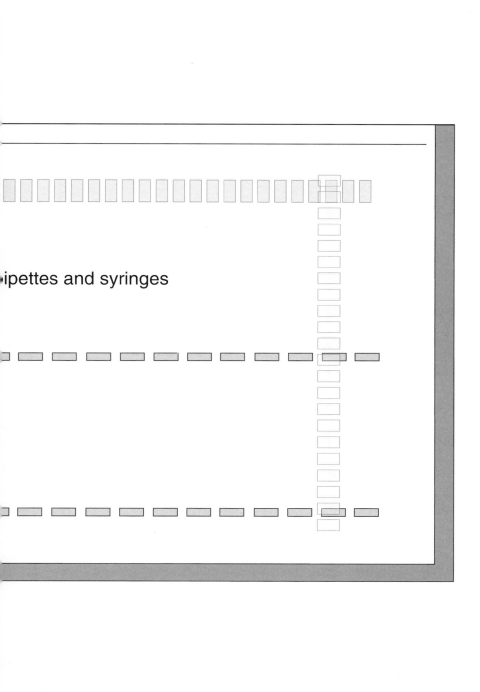

ipettes and syringes

Template for scraping out strips for IPG focusing

| 1 | 2 | 3 | 4 | 5 | 6 | 7 | 8 | 9 | 10 | 11 |

12	13	14	15	16	17	18	19	20